OurWorlds

Our Worlds

The magnetism and thrill
of planetary exploration

As described by leading planetary scientists

S. ALAN STERN

CAMBRIDGE
UNIVERSITY PRESS

PUBLISHED BY THE PRESS SYNDICATE OF THE UNIVERSITY OF CAMBRIDGE
The Pitt Building, Trumpington Street, Cambridge CB2 1RP, United Kingdom

CAMBRIDGE UNIVERSITY PRESS
The Edinburgh Building, Cambridge CB2 2RU, UK http://www.cup.cam.ac.uk
40 West 20th Street, New York, NY 10011–4211, USA http://www.cup.org
10 Stamford Road, Oakleigh, Melbourne 3166, Australia

First published 1999

Printed in the United Kingdom at the University Press, Cambridge

Typeset in Hollander 10/15pt, in 3b2 [KT]

A catalogue record for this book is available from the British Library

Library of Congress Cataloguing in Publication data

Stern, Alan, 1957–
    Our worlds: the magnetism and thrill of planetary exploration:
  as described by leading planetary scientists / Alan Stern.
        p.   cm.
    ISBN 0 521 63164 5
    ISBN 0 521 64440 2
    1. Planets – Exploration.   I. Title.
  QB601.S76   1999
  919.9′04–dc21 98-22012 CIP

ISBN 0 521 63164 5 hardback
ISBN 0 521 64440 2 paperback

For Jordan,
and his world

Contents

Preface

Modern human civilization now stretches back almost 300 generations, to the earliest organized cities. For most of that time, each clutch of humans identified their settlement and its surrounds as their home. Less than 100 generations ago, information transmission and transportation technologies were capable enough that people could form nation-states consisting of many cities and villages and consider them as a new kind of "home." In the last two generations—with the advent of space travel—many people have come to see their "home" as the whole Earth. This idea would have been essentially unthinkable to the ancients—for the world was too large for their technology to integrate the world, or even a nation-state, into an accessible and cohesive community.

So too, it is hard for us, now, to think of our "home" as being something even larger than our planet. After all, we are still trapped, both physically and to a very great degree intellectually, on our wonderful home planet. A century ago, Konstantine Tsiolkovsky described the Earth as the cradle of mankind, and that it *still* is. Yet, a logical extension of human history, and our present push outward to explore the worlds beyond Earth with robot craft, is that our descendants will very likely consider the wider realm of the solar system as much a home as we, the descendants of ancient, regional civilizations, consider the whole of planet Earth as our home today.

For a few tens of dozens of humans, those who are planetary scientists, this vision is already becoming a familiar and natural concept. And it is this concept, in part, that gave birth to the idea to bring together a few of the very best planetary scientists in the world to write about their favorite worlds, and in doing so to give a little perspective on what makes both them, and their favorite places, tick. We wanted to tell some stories of planetary exploration through the eyes of the scientists who culminate their explorations by interpreting, with human warmth, the cold 1's and 0's flashed back to Earth from sensors on spacecraft scattered across the solar system.

Of course, with nine planets and over 70 known satellites, not to mention the myriad asteroids and comets, there were many choices of locale to describe, and many fine scientists to choose from to provide personal descriptions. How to choose?

As a guiding concept, we selected scientists rather than worlds. What kind of researchers did we select? We chose individuals who had shown a deep, career-long emphasis and passion for some specific place they had been attracted to. All are known for being particularly good speakers, or writers, or both. And all were members of the second generation of planetary scientists, trained or inspired by first-generation mentors who had seen the birth of our field at the dawn of the space age. These are the intellectual 'sons and daughters,' so to speak, of legendary pioneers like Kuiper, Urey, Sagan, and Shoemaker.

We asked each of the researchers writing for this book to tell a personal story involving their own career and motivations, and to describe some part of a favorite world in which they had invested long years exploring, and to tell their story from the heart. The eight wonderful and diverse essays in this book range across the breadth of planetary science, from the inner to the outer worlds, from telescopic to robotic exploration, and from computerized armchairs to dives below the Antarctic ice. The stories encompass soaring tales of personal exploration and the dark, inner fears of a scientist living at the edge of funding difficulties. You will learn about the mountains of the Moon, the craters of Venus, the volcanos of Io, the possibility of seas on Titan, and more. Within this book you will find both a good deal of planetary science, and a perspective from several insiders about how planetary science is done. You will also see a good deal about what drives and interests planetary scientists. And, on occasion, you will see their inner hopes and aspirations revealed.

So, come and visit a few worlds across the larger home that humankind is coming to know. Come and see a little bit of the heart within our science, and the hearts of planetary scientists. Come and visit *Our Worlds*.

Alan Stern
November, 1997

Welcome home

ALAN STERN

At night, when the last rays of the Sun vanish over the western horizon, and the brilliant blue is gone from the sky, a little something wonderful happens. We planetary landlubbers can suddenly see across the deep cosmic ocean to other shores.

And what a view it is! Across the cold but crystal clear expanse of vacuum so simply named "space," shines the light of the hundred billion stars within our home galaxy. A few thousand are close enough, and bright enough, to see with unaided eyes as individual lighthouses, beaconing against the deep. The rest merely add to the faint mist of light we call the Milky Way. So too, the dark night sky reveals a few galaxies, the nearest other star islands in God's Cosmic Pacifica. But it is not the galaxies that we seek, nor even the much closer stars of the Milky Way. What calls most strongly to many of the mariners of Earth are the worlds that share the space around the Sun.

The ancients discovered the planets because these beacons moved across the night sky, changing their positions noticeably as the weeks and months slid by. And when the long eyes of the first telescopes were pointed heavenward, only four short centuries ago, they revealed that the five long-familiar wanderers of the night sky, "the planets," were wholly unlike the stars—for they appeared as rounded worlds, replete with clouds and fuzzy spots that one could imagine were mountains and valleys!

During the first ten of the twelve generations that have walked the green hills of Earth since telescopes revealed the planets as worlds, progress in understanding these places was slow. What the telescope could do well, when combined with a diligent astronomer, was chart the solar system and catalog its population. Using telescopes, planetary astronomers learned that most worlds (Mercury and Venus being the exceptions) carried with them about the Sun smaller worlds, moons, some of which were merely mountain sized, but others were large enough to be worlds unto themselves. With their telescopes, astronomers also discovered that the solar system was sprinkled with pockets of debris from the days of its origin, the asteroid belt between Jupiter's orbit and Mars, the Kuiper disk lying beyond Neptune, and the far away Oort cloud.

But the telescope and the human eye, alone, were too feeble a tool to reveal much about the worlds it could see. Looking at worlds through a telescope is something like the astronomical equivalent of trying to perform a complete medical diagnosis with a stethoscope. That didn't prevent the practitioners of astronomy from trying to picture the worlds that join Earth in orbit about the Sun, of course, but the work was hard and the results meager. So, over the first two and a half centuries of telescopic exploration of the solar system, astronomers barely learned more than how to measure a few basic attributes of each world, such as its size, mass, and the length of its day.

Of course, these first facts no more revealed the richness and wonder of the planets than would the Mona Lisa be described by saying, "image of a young woman, 120 by 70 centimeters in scale, oils on canvas." The telescope alone revealed too little.

One thing that was missing was the ability to build accurate and sensitive cameras, spectrometers, and similar devices to make a useful harvest of the light that the telescopes gathered. And so, slowly at first, but then at an exponentially increasing rate, astronomers and engineers invented tools to dissect and record the light that shone down from each world, and to thus learn something about its composition, its temperature, and its atmospheric makeup. But even with exquisite and increasingly sophisticated devices at the business end of their telescopes, even with observatories perched on mountain tops high above the worst of weather and atmospheric turbulence, the planets were still too far away to see in much detail.

What was also missing was the ability to set off from Earth and travel to the other worlds. To see these places up close. To map them carefully. To land on their surfaces, and probe their atmospheres. To touch and feel them. To make them real. To leave astronomy behind for a new kind of science, a mix somewhere between astrophysics and geophysics, called planetary science. To come to know worlds by going to them. To be there.

Cradle vista

Whoever it was that said that travel is broadening certainly had it right. The single most humbling lesson we have learned in planetary exploration is just how incorrect (and usually naive) our first astronomically based perceptions of the planets were.

Why was this the case? For one thing, telescopic observations generally produce too low a resolution to really see the details of the worlds we study. Telescopic observations are also fundamentally limited in their ability to reveal the compositional and physical details that *in situ* measurements on a planet, such as wind speed, mass spectroscopy, gravitational harmonics, and seismic studies, can.

In essence, our ability to understand the planets from Earth was about equivalent to trying to fully understand the geology, climate, and cultures of, say, Asia, by flying over in a space Shuttle with a pair of binoculars. As a result, most of our conceptions about the planets, before the age of space exploration, were, ... well, *mis*conceptions.

Remember? We thought Mercury had an atmosphere, with clouds, and Venus might be an inhabited swampland. Mars seemed to be green with vegetation each spring. And the asteroids were thought to be the remnants of an exploded planet. The giant planets were uninteresting, except for that funny red spot on Jupiter, and only Saturn had rings. The moons of the giant planets were boring, and pretty much all the same.

Today we know that *none* of these ideas, all based on the best available evidence at the dawn of the space age, just 40 years ago, was correct.

To learn what the planets were truly like we had to go there, but it wasn't easy. The first impediment was distance. Even with our best rockets, crossing the great, vacuous gulf between the planets takes months within the compact inner solar system, and years in the outback beyond Mars and the main asteroid belt. The second impediment was money, for the tools of technology are expensive. Even in the 1990s age of smaller, cheaper missions, individual spacecraft often cost more than small fleets of jet transports, and their launch vehicles usually about double that expense. As Tom Wolfe wrote "No bucks, no Buck Rogers." The third impediment was (and to a large extent still is) the heavy engineering required to make spacecraft and their launch vehicles work reliably. Space pioneer Werner von Braun once said that, "We can beat gravity, but the paperwork is enormous."

Nevertheless, the political necessities of the cold war forged a pathway to the planets in the form of a very public competition between the United States and the Soviet Union. What were the two superpowers trying to prove? That each had a society that could lead the world. And lead they did, for the historic explorations they financed and executed will stand, as long as humans record their history, as positive testaments to the United States and the Soviet Union, and to the curiosity, prowess, and ingenuity of twentieth-century civilization.

Free bird

And so, owing to politics (as opposed to manifest density, or greed, or even very much to scientific curiosity), beginning in 1959, we went exploring.

Our first steps were simply aimed at the Moon, less than one hundredth the distance to the nearest planet, for it was a time for learning the basics of how to launch and fly spacecraft. Launch vehicles exploded. Rockets went off course. Spacecraft spun out of control, lost radio contact, or failed for one of a dozen other reasons. Soviet records are sketchy, but US records show that, of the first four Pioneer and six Ranger missions to the Moon in the late 1950s and early 1960s, *none* fully succeeded, and only three of these ten missions could be called a partial scientific success.

The first coup in planetary exploration came early, in 1959, when the Soviets launched Luna III on a week-long mission to obtain the first images of the heretofore-hidden far side of the Moon. Luna III's main scientific result was in discovering that the lunar far side has a vastly different appearance from the front side. This, like many early discoveries that were to come, was a lesson primarily about how little we had known, or could know, from studying far away worlds solely from Earth.

By the early 1960s, both the United States and the USSR had undertaken vigorous programs to make flybys of the two nearest planets, Venus and Mars. Only about half of the attempts succeeded; but when they did, they spoke volumes. Later in the 1960s, while the robots made further forays to Earth's two nearest planetary neighbors, Apollo spacecraft delivered nine human crews to the vicinity of the Moon. Six of these crews were sent to the surface to deploy instruments, collect samples, and explore the geology of lunar mountains, valleys, and plains. It's a shame, but human exploration paused there, never to be restarted in the twentieth century.

The 1970s saw Venus and Mars exploration move into a more sophisticated phase, with entry probes, landers, and globe-circling orbiters designed to make far more in-depth studies than simple flyby visits ever could. And, later in the 1970s, the United States branched further afield, launching a triple flyby mission to Mercury and four different flyby spacecraft to Jupiter.

Few US missions were launched in the 1980s. However, the Russians continued their spectacular string of successes with Venus exploration, and the three of the US probes that reached Jupiter in the 1970s went on to reconnoiter Saturn. One of these probes, Voyager 2, was even sent on to make the historic first explorations of the Uranus and Neptune system. Meanwhile, a flotilla of European, Russian, and Japanese flyby missions was launched at Halley's Comet, and a US spacecraft was redirected to fly by another comet, called Giacobini-Zinner. By the time the 1980s ended, all of the planets save Pluto had been visited, and the Galileo mission was *en route* to make the first close reconnaissance of an asteroid, and then on to orbit Jupiter.

And what of recent times? The 1990s have witnessed a reflowering of planetary exploration, with many more launches occurring than in any decade than since the 1960s. Many of these missions were smaller and more focused than the expensive, "do-all" explorations conceived in the 1970s. It's a good thing too, because the larger missions had grown in complexity and cost to a point that they were not financially sustainable.

Today, as this book is being completed, we have just witnessed the highly successful, low-cost Pathfinder Mars landing, and the NEAR mission is approaching one of the very largest asteroids in near-Earth space for a detailed orbital mapping mission. So too, missions are either underway or being built to orbit and land on comets, return samples from asteroids and comets, put long-duration rovers on Mars, and explore the multifaceted Saturnian system. There is even talk of returning humans to the fray, with missions to the Moon, nearby asteroids, and Mars envisioned not so long after the century turns. Planetary exploration is alive!

21

As we stand at the crossroads of time between two centuries, and indeed, between two millennia, we also stand at a kind of crossroads in the history of our species, and of our planet. Mother Earth has produced tens of millions of species, one of which, now, is taking its first tentative steps away from its birthplace. As we *homo sapiens* follow that course, in part, at least, for reasons we ourselves do not fully understand, we are coming to see the larger venue of the solar system as an identity associated with our own. We are slowly—but surely—gaining a planetary perspective beyond the cradle of Earth.

The solar system is a vast place, but perhaps no vaster for us now than the Earth was to the ancients. We planetary scientists have the pleasure and the privilege of being on the vanguard of the first, primitive wave of exploration across the sea of space and to the other shores warmed by the Sun. Come and see a little of what it is like to participate in this exploration, as we invite you into our home.

Inner Worlds

1

Exploring Mars

STEVE SQUYRES

Steve Squyres is a man in motion, a man of constant energy, enthusiasm, and scientific verve. Few planetary scientists can keep up with his pace, or his intellect. In addition to being a professor at Cornell, Steve serves on countless NASA (National Aeronautics and Space Administration) committees, participates in numerous space missions, keeps up a research pace few can match, and with his wife of many years manages to raise a beautiful family. In the essay that follows, Steve takes us on a journey to Mars, via Earth.

When I was a kid, I loved maps. Still do. I grew up in the 1960s, and in those days if you looked at a not-so-current atlas of the world, you could still find a few blank spots—places that were understood poorly enough by the map makers that they had to confess to not quite knowing what to draw there. Somehow, I loved the idea of an incomplete map, with places on it still to be discovered. As a boy I read everything I could on exploration—Amundsen and Scott in the Antarctic, Beebe and Barton in the deep sea—tracing their exploits across my maps and dreaming of the exploring I'd do myself someday.

By college, things felt less dreamy. The blank spots on the maps were gone. I had originally picked geology as a major because it might allow me to combine my two loves of science and mountaineering, and I was drifting toward something involving the geology of the sea floor—at least there were still some blank spots there. But it wasn't clicking.

Then, early in the spring of my junior year in college, I was giving my fiancée a tour of campus. This was 1977, shortly after the Viking spacecraft had arrived at Mars, and, while we were in the Space Sciences building, she spotted a little 3×5 card tacked to a bulletin board announcing that a professor who was a member of the Viking Orbiter Imaging Team was going to be teaching a graduate seminar course on Mars. What the hell, I thought, and I went.

It didn't begin well. In fact, I nearly got kicked out of the class. First thing the prof asked us was "Are there any undergraduates here?" One timid hand went up (mine), and he told me to see him after the lecture. I could tell from the look on his face he was going to give me the boot. What saved me was that one of my would-be classmates was a grad student from geology who knew both me and the prof, and who came over at the crucial moment and vouched semi-truthfully for my studious nature. The prof assented, and explained quite reasonably why he preferred not to have undergraduates in a course like this one. If, say, he was lecturing about martian temperatures, he said he wouldn't want to have to stop and explain a concept as basic as thermal diffusivity. I just nodded wisely ... and then dashed back to my dorm room to look up what thermal diffusivity meant.

Because it was a graduate course, we were expected to do a little original research. A few weeks into the semester I figured I'd better start thinking about my first in-class presentation, and got a key to the "Mars Room," where all of the brand-new pictures from the Viking orbiters were being kept. I will not forget that day. Some of the pictures were in shiny blue three-ring binders; most were still on long rolls of photographic paper, stacked on the floor or in their shipping cartons. My idea was to spend fifteen or twenty minutes flipping through pictures in search of a paper topic. Instead, I was there for four hours, and what I saw in the pictures stunned me. I understood none of it, of course, but that was the beauty of it. *Nobody* understood some of this stuff; in fact, only a handful of people in the world had even had the chance to see it yet. Sitting there cross-legged on the linoleum, cardboard boxes stacked around me, I was exploring a new world. It was a revelation.

As the weeks passed and I stared at the pictures more and more, I began to think that I understood some of what I was seeing. For my first research project, I spotted some little dried-up valleys near the equator of Mars that opened out into a shallow basin, and suggested to the class that maybe a small martian lake had existed there, billions of years ago. Not particularly profound, but it's an idea I pursue to this day. The reason I pursue it, though I could not have articulated it well then, is tied to martian climate, and maybe even to martian life.

Mars approaches

Among all the planets, Mars has always held a special place in the human imagination. The reason for this is that it is more like home than any other place in the solar system. It has a day a little more than 24 hours long. It has seasons. It has volcanos and dry riverbeds, pastel skies and clouds. It is a world we can imagine inhabiting, or being inhabited by others.

The idea that Mars could harbor life has been a remarkably resilient one. It goes back more than a century, to the work of the Italian astronomer Schiaparelli, and later of the American Percival Lowell. Schiaparelli observed Mars through his telescope, and noted what appeared to be a web-like network of straight lines lacing the surface. He called these lines *canali*, or channels, a term that was unfortunately mis-translated into English as "canals." Lowell carried on later with Schiaparelli's work, and took the bold leap of suggesting (as Schiaparelli had not) that the canals were the intentional creations of intelligent martian life forms. Lowell made intricate maps of the canals, and argued that their remarkably regular geometry could only be the product of intelligence. He was right, of course; unfortunately, the intelligence was at the wrong end of the telescope. The canals were nothing more than optical illusions, created when the human observer's eye and brain erroneously connected faint dark markings on the martian surface with straight lines.

As observing technology improved through the 1930s, 1940s, and 1950s, the illusory nature of the canals became clear, and the idea of intelligent martian life waned. But in its place arose a new hypothesis. Though high-resolution photos taken by the world's great telescopes showed no canals, they did reveal patterns of dark markings on the planet that changed with the seasons: a "wave of darkening" that progressed cyclically across the planet's surface, as would be the case for the seasonal waxing and waning of vegetation. So the concept of life on Mars went from little green men to lichens, mosses, and cold-resistant cacti.

Even this modest picture was dashed in the 1960s with the advent of space exploration. The first simple Mars probes, Mariners 4, 6, and 7, flew by the planet and showed a cold, hostile world with craters like the Moon's and no hint of life. Scientists began to suspect, correctly, that the wave of darkening was the result of nothing more than the cold martian wind, blowing bright dust and dark sand to and fro across a lifeless surface.

Despite the discouraging results from the three Mariner flybys, one more set of missions was planned: Mariners 8 and 9, the first planetary orbiters. Mariner 8 was launched on a warm Florida day in early May of 1971. Liftoff was flawless, but within minutes both launch vehicle and spacecraft lay in pieces on the floor of the Atlantic, victims of a flaw in the Atlas rocket's guidance control system. A furious scramble to pinpoint the error before the end of the month-long launch window ensued, and three weeks later the surviving Mariner 9 spacecraft was put successfully on a trajectory to Mars.

Mariner 9's cruise went smoothly. Then, just a few months before its scheduled orbit insertion, telescopic observers on Earth noticed a dust storm beginning to build. The storm grew to massive proportions, eventually enveloping the entire planet. By the time the spacecraft had completed its journey and settled into orbit around Mars, the planet's surface was completely hidden from view.

The storm raged for weeks. Mariner 9 had only been designed with three months of mapping in mind, and the scientists were in agony as the days slipped away. Fortunately, the engineers at the Jet Propulsion Laboratory who had built the spacecraft had been conservative—very conservative—in their projections, and the spacecraft survived over a year in orbit, more than four times its design lifetime. As the dust finally started to settle, several dark spots on the planet's surface began to come into view. The Mariner scientists watched day after day in amazement, gradually realizing that these spots were actually the summits of huge volcanos, emerging from the subsiding dust cloud like islands from an opaque sea. Slowly the planet was revealed to Mariner's cameras in all its grandeur.

Mariner's discoveries were astonishing. They included:

- huge volcanos, the largest of which, Olympus Mons, rises 27 km from its base to its summit, more than three times the height of Mt Everest;

- a giant canyon, up to 6 km deep and stretching along the equator for some 5,000 km, the distance from New York to Los Angeles;
- finely layered deposits of dust and ice at the north and south poles, preserving an intricate record of climate changes on the planet;
- enormous channels, cut by floods that rampaged across the martian surface with the power of a hundred Amazon rivers.

Among these wonders, the most important of Mariner 9's discoveries were some of the least impressive: the planet's small valley systems (see Figure 1.1, color section). These are not spectacular features by any means. They are simple dendritic channels, much like dry riverbeds on Earth. A few hundred kilometers long at most, typically 1 km wide, their scenery would merit nothing more than a nice state park on this planet. But in their mundane nature lies their true significance: there is no way to make features like the valley systems on Mars today. Unlike the great flood channels of Mars, the valley systems were created by a comparative trickle of water. A trickle of water would freeze if the weather were as cold as it is on Mars today, and would freeze quickly. These valleys, then, told scientists something that nothing else in the Mariner images did: there was a time in Mars' past when the planet was warmer, wetter, and more like Earth than it is now.

That time was long ago, because the valley systems of Mars are old. In fact, most of them date from the earliest epoch of martian history, some 4 billion years ago. The significance of this is simple, and it hit the Mariner scientists between the eyes: *the warmer, wetter epoch of martian history coincided with the same period in which life first came into being on Earth.* And so, the obvious question arose: if it happened here, could it have happened there? Gone were the dreams of green men and lichens, but in their place arose the first real scientific basis for believing that life could have once existed on Mars.

The next mission to Mars was Viking, an ambitious attempt to address that possibility. Viking consisted of two orbiters and two landers, and this time all of them worked. The landers were designed primarily to test one simple hypothesis: that microorganisms live today in the soils of Mars. The first of the Vikings touched down on the plains of Chryse Planitia on July 20, 1976, and the second landed on Utopia

Planitia six weeks later. Each lander carried a camera, an arm to scoop up the red martian soil, and a small battery of biology experiments to search for signs of life in it.

The search failed—the scientific consensus was that the evidence Viking found regarding life was negative. But almost lost amidst the hubbub surrounding the landers' vain search for microbes was the wealth of data returned by the orbiters. Designed primarily to scout out landing sites, these orbiters too survived long beyond their design lifetimes, and imaged the planet again in beautiful detail. The Viking landers failed to find evidence for living organisms in the soil, but the orbiters' images solidified the case for a warmer, wetter early Mars.

Ice world, Antarctica

After Viking, the spotlight in planetary science shifted away from Mars for a while. The two Voyager spacecraft were launched in the fall of 1977, beginning their spectacular Grand Tour of the outer solar system. I was fortunate enough to get involved with the Voyager imaging science team during grad school, and went on to write a thesis about two of the moons of Jupiter. After all, if you're an ambitious young lad there's no substitute for working on the hot topic of the day. But I didn't forget about Mars, or about its ancient valleys and lakebeds. I still didn't know what they meant. How warm and wet had Mars really been? Like a jungle? Probably not. Like the Sahara? Like Antarctica? It was a puzzle, and, I still felt, an important one.

After grad school I landed a job at Ames Research Center, a first-rate NASA facility in the heart of Silicon Valley. Ames was a dream for me. Besides earning a real salary for the first time in my life, I also found myself surrounded by some of the top scientists in my field. Among them were Chris McKay, Bob Wharton, and Dale Anderson, who I quickly learned shared my passion for understanding Mars. And not only were they doing something about that passion, they were doing something about it in Antarctica. This was an opportunity I couldn't pass up.

There are many reasons to make field work on Earth part of our exploration of the planets. One is that it's fun, if the truth must be told.

Since antiquity the urge to have a great time has been behind many an explorer's schemings, and sometimes the urge is thinly veiled at best. But field work, if done under the right circumstances, can also provide a reality check on our wildest theories and our sketchiest data. You may think you're clever enough to look at martian images and spot evidence of an ancient environment where life once could have thrived. But if that environment has an analog on Earth, and if that analog is barren and dead, you'd better think again.

The analog we were interested in lay in a strange corner of Antarctica. The continent is a vast frozen desert. Most of it is covered by the Antarctic ice sheet, but a small piece of it is cut off from the ice by the bulk of the Transantarctic Mountains. This region, known as the Dry Valleys, is the most Mars-like place on Earth. There are no living animals there, no insects. Not a blade of grass grows. Even the hardiest of mosses and lichens are almost nowhere to be seen. It's a land of rock and sand, dust and wind. The average annual temperature is 25 degrees below zero—only a few tens of degrees warmer than Mars is today.

Despite the harshness of the Dry Valleys, there are lakes there. Tropical lagoons they are not. Each has a year-round cover of ice some 3 to 6 meters thick, but beneath the ice lies liquid water. If ever the valley systems of Mars had led to the formation of lakes, these ice-covered lakes in the Dry Valleys were the closest thing to them that we could find on Earth.

Our goal was to explore the lakes, and to see what clues they might hold concerning early Mars. We knew from the pioneering work of others, notably George Simmons from Virginia Tech, that the lakes contained simple, primitive life, in the form of mats of algae, similar to some of the earliest organisms that lived on Earth. We wanted to learn more.

Antarctica can be a difficult place to work. The cold is what gets your attention at first, but that's actually not particularly hard to deal with. The bigger problem is the isolation. Although the US Navy does a superb job of supporting field parties via helicopters ("helos" in Navy parlance) from their base at McMurdo Station, the work has its frustrations. Equipment breaks in the cold and needs to be replaced. Storms play havoc with helo schedules. All in all, it's a better-than-passing analog not just for the physical conditions of Mars, but also for some of the challenges humans will face working there. The problems, though, are more than offset by the stark beauty of the place, and by the simple joy that comes from doing what you love. Our field camp lay near Lake Hoare in Taylor Valley, at 78° south latitude, and I tried to describe some of what we were experiencing there in letters home to my wife:

> Camp consists of one main building, which is actually what's usually called a Jamesway: sort of a canvas Quonset hut stretched over a wooden frame. This serves as a kitchen, dining room, dive locker, meeting room, etc. (It's called the Hoare House; don't blame me.) Then there's a john (pretty basic), a generator shack (ditto), and another small building that serves as a lab. We sleep in two-man mountaineering tents. The rest of camp is filled out by an unattractive assortment of 55-gallon drums of diesel and helicopter fuel, discarded crates, an air compressor, and scuba tanks. It is definitely not the most beautiful facility you've ever seen. However, it is in one of the most beautiful locations you could imagine. The valley is narrow, and the peaks rise to 6000 feet above the valley floor. Blocking off the valley and damming the lake is the Canada Glacier. All along the glacier terminus is a very imposing 100-foot ice cliff. The land around the lake is completely barren of life—very Mars-like. All in all, it gives a sense of magnificent desolation.

The work began several days later with what amounted to a practice dive into the waters of Lake Hoare (see Figure 1.2, color section):

> Today was my first dive. Setup is like dressing a six-foot kindergartner to go out and play in the snow. First comes the polypropylene, which one week into the trip is already starting to smell like a damp reindeer. Then the fancy blue divewear. At that point, I start feeling warm. Then the blue dive booties. Then the drysuit—from that point on, I can't dress myself. Then velcro'ed ankle weights. Then duct tape over the ankle weights. Then rubber-soled boots so I can waddle out to the dive hole. My two buddies on the support crew drag the tanks, weights, and other gear out on a sledge. Then the boots come off. Then a heavy chest harness with a nylon tether goes on

with straps and carabiners. Then a weight belt with 40 pounds of lead, then the tank and backup air (in what's called a pony bottle). Then the mittens. Then the hood goes up and the mask is put on. Then the comm check—I just had them crank the volume all the way up, because it's really hard to hear over the bubbles. At that point, I'm ready to boogie. The hole's about 4 feet in diameter, and the ice is about 12 feet thick. I edged into the water and bobbed like a cork. Not enough lead. Back out, strip to just the drysuit, put on more lead, and do it again. This time, everything was cool. I worked my way down the hole slowly, taking it all in. The ice was a beautiful light blue, with intricate columns of bubbles in it, like carved crystal.

The hole for this dive was the one called the glacier hole—right up against the face of the glacier. It was dark down there, but lighter than, say, a night with a full moon. The water was clear—at 65 feet, you could look up and see the dive hole plain as day (it looked very reassuring)—and very still. The bottom was sandy and hummocky. The amazing thing was the glacier, which underwater looked bigger, somehow, and whiter, and more menacing than any I've seen before. I swam over to it, and climbed around on the face for awhile, just using my hands. Moving around was no trouble, and except for my hands I was at least as warm as in a rented wetsuit in Monterey Bay. The time passed amazingly quickly. I think looking upwards on the ascent was the best part of all. Everything was a beautiful, ghostly pale blue, and my bubbles trapped against the bottom of the ice looked like inverted puddles of mercury. Went up through the hole and was hauled out like a very big fish, and it was all over.

A few days later the science began:

Today was the second dive. This was in dive hole one, where the water is just 30–35 feet deep. The bottom was mostly sandy, with a steep slope. Most of it was covered with the algal mat, which looked like an enormous olive-green army blanket lying on the sand. In places, the photosynthesis had produced enough oxygen (and hence enough buoyancy) that it was just on the verge of lifting up—ooch-ing up off the lake bottom as if draped over a bunch of doorknobs. As I walked across it, whole segments, some several feet across, broke free and started drifting up to the surface, looking like tat-tered laundry blowing in the wind in slow motion. Fantastic.

The bad news is that our second compressor broke today, leaving us with none. It's a lousy design, and the thing basically vibrated so much it shook itself apart. They're going to fly it back to McMurdo tomorrow, and with luck we'll have it fixed and back out here tomorrow evening. We'll see.

Things didn't always go well, as this dispatch a week or two later shows:

Damn damn damn damn damn. Bad day today. First, no resupply helo—we're down to not much more than PB&J [peanut butter and jelly]. Second, Dale had a monster nosebleed while suiting up. We're talking gusher. Then, to top it off, he tried to go down dive hole 2 to pull the sediment traps there before I take some cores, and found the hole wasn't completely melted through. The melter had gotten most of the way down and then slipped sideways, creating a nasty wedge-shaped shelf at the bottom. The shelf was too thick to chop through; Dale tried valiantly, but wound up accomplishing nothing more than dropping his chopping tool down the hole. So now we've got the coil back down the hole and are melting again, having lost a full day screwing around. This we need like a hole in the head. Ah well, tomorrow may be a better day.

It wasn't:

Just awoke to the sound of helo blades. Yowza. Turns out, though, to be a load of Kiwi "distinguished visitors" on a tour, showing up unannounced. They brought no food, and now we're down to noth-ing at all but peanut butter and jelly, mandarin oranges, and fro-zen lobster.

A few days later:

When it dumps on you, it dumps on you. I was suited up yesterday, waiting for Dale to pump up my tank, and we blew out a gasket on the latest compressor. I had to strip back down, help take the thing apart, and then try to come up with a way to fix it. We glued the breaks together with epoxy, coated both sides with teflon tape, and pressed them between two aluminum dinner plates with scuba weights. We let the thing set overnight, and with great anticipation put it back together and fired it up today. Last time, the gasket had blown at 800 PSI. It crept up to 800 . . . 900 . . . and then the ac- cursed thing shook itself apart at another *fitting.*

On the plus side, I took a nice hike today up the lake to look at car- bonate deposition. Interesting science, and a nice walk, though brisk. I just got back into camp. Dale's working on the compressor joint, Dean's cooking dinner (lobster with mandarin orange sauce, and peanut butter and jelly for dessert; big surprise there). Rumor has it we might get a helo in two days. I hope it brings food, mail, and beer . . . and drops the compressor into McMurdo Sound.

Eventually things got back on track:

Things have kicked into high gear again. We got another surprise helo a few days ago, and it brought us the fittings to fix the busted tube on the compressor. The gasket fix worked, and we're back in business. Yesterday morning I got in a terrific dive in hole two. I laid out a 3 × 3 meter grid of stakes and nylon lines, defining nine points at which to take cores. It involved lots of hammering, and by the end the visibility was about a foot. Then I did a little under- water geologizing, and found a kind of sedimentary structure that I've never seen or heard of before. The bottom of the lake is covered with small mounds, maybe 50–100 cm across and 20–30 cm high. We think they're formed when sand is released from local sources in the ice, spreading out slightly as it descends through the water col- umn and forming a little mound on the lake bottom. If we ever see something like this with a rover on Mars, I'll think I know what I'm looking at (and, Mars being Mars, I'll probably be wrong).

The cores, by the way, are fabulous. For one thing, they're full of layers of algae. Also, there are two beautiful carbonate layers in each core that make excellent marker beds. But the position of the carbonate layers varies hugely from one core to the next, even over a distance of just 1.5 m. This, of course, provides additional support for our idea of how the sand mounds form. And it's one more thing to look for on Mars someday.

We'd been out in the field quite a few weeks by this point, so the brightest moments tended to involve life's simpler pleasures:

NEWS FLASH: Helo this morning brought FOOD: Real stuff . . . T-bone steaks, pork chops, corn, peas, carrots, fresh peppers, bananas etc., etc., etc. We're in heaven!!!

We wrapped up the work at Lake Hoare, and prepared to move up the valley to Lake Joyce:

4 AM . . . so tired. Have spent the last 13 hours getting ready for the move to Joyce. Helos come in five hours. Gotta get a little sleep soon. Heavy work, dragging around the 200-lb generators, 300-lb compressor, 400-lb kerosene drums, and the heater for the glycol, which I swear must weigh about 25 tons . . . I think it's made of the same stuff neutron stars are made of. Ice surface is incredibly rough now—3 feet of relief, lots of sand, ponds of water. Had to drag stuff 100's of yards in "banana boats"—round-bottomed sledges. As Dean put it, to work on this project you have to be "strong as a bull, smart as a tractor." Lots more to write about, but too tired. Ready to go, but weather's going to hell. We'll see.

The letters end there. Next came the camp move to Lake Joyce, setup there, and five days of round-the-clock melting of a dive hole without hitting water. As luck would have it, I had a helicopter to McMurdo and a flight home scheduled after five days, and had to take it. They broke through into the lake the day after I left.

I learned a thing or two at Lake Hoare. One thing I picked up were some ideas, simplistic though they may be, of what to look for in the geologic record when we finally do sample an ancient martian lakebed. The wonderful thing about the cores we took was that they contained stromatolites: layered structures of algal mat and sediments. Stromatolites are rarely found, but fossil stromatolites more than three billion years old provide some of the first evidence of life on Earth. Direct ancestors of the primitive organisms in Lake Hoare existed on Earth at the same time we think lakes existed on Mars. Could similar martian organisms have existed too, and could they be preserved in the martian geologic record?

Another thing I learned at Lake Hoare was just how hard it can be to work in a really remote environment, with balky equipment and tired humans. When the time comes, it isn't going to be any easier on Mars.

Childhood's end

Around the time I was doing the work at Lake Hoare, something about martian water had begun to bother me. Where had it all gone? It had required a fair amount of water to carve the small ancient valleys, and a whole lot more to carve the great flood channels. So where was it? Some was definitely locked up in the polar deposits, but not much. A bit was trapped in minerals in the soil, and a bit more had escaped to space. But from what I could see, the numbers didn't add up. More had to hiding someplace. The logical place for it to be was frozen in the ground.

The word "permafrost" is often misused. It doesn't mean that there's ice in the ground; it only means that the ground is colder than the freezing point. But most permafrost on Earth does indeed contain ice, and it was clear that there was a lot of permafrost on Mars. Is that where the water was hiding, locked up as ground ice?

In my days as a geologist, I had learned about features called "rock glaciers"—strange accumulations of rock and ice, common in some alpine regions, that are able to deform and flow because of the ice they contain. In essence, they look like piles of rock, but they flow like glaciers. I had also begun to notice, in all of my staring at pictures of Mars, that something strange began to appear when you looked at the middle to high latitudes on the planet. Instead of being sharp and angular, most of the slopes there seemed soft and rounded. I mentioned this to Mike Carr of the US Geological Survey, one of the true legends of the Mars game. Mike said he had seen the same thing. Could it be because there was ice beneath the ground there, and the ice was allowing the soil to deform and flow? Or was it something else?

Mike and I decided to try to test our impressions by mapping this "terrain softening" over the entire planet, looking at some 50,000 Viking orbiter images. Sure enough, the equatorial latitudes were crisp and sharp everywhere. But past about 30 degrees, flow was evident almost everywhere there was enough topography to show it.

By itself this observation proved nothing, but it was one piece in the puzzle. Another was provided by some elegant work done that same year by Fraser Fanale of the University of Hawaii. Fraser took a theoretical look at the problem, and found that if you started with the martian soil rich in ice everywhere and waited four billion years or so, sure enough you'd lose the near-surface ice from latitudes lower than about 30 degrees, but keep it (and actually accumulate ice) at higher latitudes. So, we had a consistent story on where the ice seemed to be. Problem is, a consistent story is not the same thing as data.

Where the ice is matters for more than just academic reasons; when humans finally go to Mars they're going to *need* ice to make drinking water and rocket fuel. Observations hinted that the ice was there, and calculations suggested it ought to be present within less than a meter of the surface once you got past latitudes higher than about 60 degrees. Surely, it seemed, there should be some way to confirm these ideas … but it would take another mission to Mars.

By the mid-1980s, Mars exploration began to get back in gear. The next mission planned was Mars Observer (MO). MO was to be another orbiter, but one that was a quantum leap beyond Viking. In the years since Viking had flown, enormous strides had been made in instrument technology. Where Viking had carried TV-like cameras and a few other very rudimentary remote-sensing instruments, the technology of 1985 allowed a whole new class of advanced sensors to be sent: infrared spectrometers to map rock mineralogy, altimeters to map topography, sophisticated new meteorological sensors, and more.

The idea behind MO seemed like a good one: take an existing and tested spacecraft design, make the modifications needed for it to work at Mars, load it up with the best instruments you can build, and revolutionize our understanding of the planet in the process. The spacecraft chosen was an Earth-orbiting satellite built by RCA. To get it to work at Mars they gave it bigger solar panels, changed the thermal design—did all the things you need to do to a spacecraft to get it to work farther from the Sun. Or at least they thought they had.

To pick the instruments and the science team for MO, NASA put out what's called an Announcement of Opportunity, or AO. The rules are elegantly simple: anyone can propose, and the best proposals win. Reality is of course a bit more complicated than that but, even so, the process really is remarkably open and fair.

The Mars Observer AO came out in May of 1985. At that point, less than four years out of grad school, I didn't feel I was in a position to write a proposal for an instrument to go on a spacecraft. Instrument proposals usually require large teams and years of preparation and experience, none of which I could offer. But I also read that NASA was planning on providing two "facility" instruments for scientists to use. One was an infrared spectrometer, and the other was a gamma-ray spectrometer (GRS). These, I saw, were my opportunity to get started in the space exploration game for real.

Infrared spectroscopy I understood, at least a little. Gamma rays were a mystery. But the one thing I did know about gamma-ray spectroscopy at Mars was that it could detect ice beneath the surface. Gamma rays are photons of enormous energy, roughly a billion times the energy of visible light, and they can penetrate through many centimeters of rock or soil. Gamma rays produced in ice as much as a meter below the martian surface could make it to space and be detected from orbit. It was the best technique that MO offered for finding out where some of Mars' water had gone, and I decided to give it a whirl.

I didn't find it easy to try to teach myself nuclear physics, but I had no choice. Gamma rays are produced within the nuclei of atoms, and if you want to write a decent gamma-ray proposal, you'd better have your nuclear physics straight, at least more or less. First I bought myself a nuclear physics text, and I read it, cover to cover. Then I found every paper that had ever been written on planetary gamma-ray spectroscopy, and read all of them, too. By this time I knew just enough to be able to start asking dumb questions; the thing I really did right was direct my questions at Jack Trombka, of NASA's Goddard Space Flight Center and one of the true wizards in the esoteric field of planetary gamma-ray spectroscopy. Jack had nothing to gain by answering my questions, but he answered them anyway—clearly and with considerable patience. It took me eight months to write a ten-page proposal, but when the call from NASA Headquarters came, I was on the team.

Mars Observer was an exhilarating experience while it lasted, and it really lasted quite a while. The team was formed in 1986, and the space-craft didn't launch until 1992. The intervening years involved a lot of work for me, and far more for most of my colleagues on the project. There were gamma-ray science team meetings in Arizona to guide the design and construction of the instrument, and MO Project meetings at the Jet Propulsion Laboratory (JPL) to argue the fine details of how all the instruments would work together to map the planet. There were even two more trips to Antarctica, flying a gamma-ray spectrometer on a stratospheric balloon there to simulate production of gamma rays on Mars. It meant a lot of time away from home, but I was doing what I had always wanted to do—participating, finally, in a real mission to the planet that had captivated me for a decade. The moments I remember most vividly came when I visited the facility where the spacecraft was built, in East Windsor, New Jersey, just a few weeks before it was shipped to the Cape for launch. Perched in the blackness of the thermal vacuum test chamber, it looked like an enormous jewel (see Figure 1.3, color section).

MO never made it. It thundered beautifully into space atop a Titan 3 rocket in September of 1992. Cruise was uneventful, and our GRS checked out perfectly. Eleven months later, I traveled to JPL to view the data from the orbit insertion. I checked in to the Pasadena Holiday Inn 72 hours before the burn, and flipped on the TV to catch the late news. The first words I heard were chilling: "Scientists at the Jet Propulsion Laboratory are struggling tonight to regain contact with the Mars Observer spacecraft ..." A quick phone call to the lab confirmed it: transmissions from the spacecraft had stopped a few hours before.

Three days before igniting the big rocket motor on the spacecraft to ease it into orbit, the spacecraft's propulsion system had to be readied for the burn. This involves opening a "pyro" valve (using a tiny controlled explosion), and letting high-pressure helium flow into the fuel and oxidizer tanks. Pyro valves are remarkably reliable devices, but firing them poses a slight danger to some sensitive components of the spacecraft, like the radio transmitter, if they are on when the pyro is fired. The sequence that MO was supposed to execute was to turn off its transmitter, fire the pyro valve, and turn the transmitter back on. The transmitter turned off at the appointed time, and we never heard from the spacecraft again.

We don't know what happened to Mars Observer, and we never will. There are theories, one of them being that a flaw in the design of the propulsion system (brought about in part by the spacecraft's transition from an Earth-orbiter to a deep space vehicle) allowed a little fuel and oxidizer to mix in a pipe once the valve was opened. They would have reacted violently, burning through the pipe and venting propellants into space. There are other theories too. The only thing I knew was that Mars Observer had died, and that part of my dream of exploring Mars had died with it.

Mars alive?

The loss of Mars Observer affected me deeply and personally—more than it should have, probably, but there you have it. There was no way to bring the spacecraft back, so instead I threw myself into finding a way to replace it. Inquiries were conducted, advisory panels were formed; I let myself become consumed by participation in them. When it came time to present MO's replacement to the Congress, I organized letter-writing campaigns and traipsed endlessly through the corridors of Capitol Hill, an obvious misfit among the well-heeled lobbyists with their tailored suits and cell phones.

The replacement for MO that came out of all this activity didn't look much like MO. The Mars Observer instruments were first-rate, and it was clear that we needed to find a way to get copies of them back to Mars. But flying them on a copy of MO wasn't going to cut it. For one thing, it couldn't be afforded—just the huge Titan launch vehicle that had sent MO to Mars had cost more than $100 million. For another thing, a copy of MO was inconsistent with the new "faster, smaller, cheaper" philosophy of space missions that had taken hold at NASA by 1993. This phrase, repeated by NASA managers like a mantra, encapsulated a simple truth: if all your missions are billion-dollar extravaganzas, you're not going to fly very many missions.

The replacement for MO was the Mars Surveyor program. Unlike MO, Mars Surveyor is a *sequence* of missions: small, lightweight, and comparatively inexpensive spacecraft launched to Mars every 26 months—the length of time it takes Mars and Earth to realign in the proper configuration for a launch from here to there. I believed it to be a good concept (and I still do), and I fought hard for it on Capitol Hill. We won that fight, and as I write this in the fall of 1997, the first of the Mars Surveyor spacecraft has just gone into orbit around Mars, carrying copies of five of the original MO instruments. It looks poised to provide a flood of new and fantastic data.

The bittersweet irony for me in all of this is that my beloved gamma-ray spectrometer is not on the spacecraft. If you use a smaller spacecraft and a smaller rocket, some of the instruments have to stay home. The sensible thing for NASA to do was to fly the lightest instruments early and the few heavier ones later, so as to get as much science done as quickly as possible. The GRS was, alas, one of the two heaviest instruments on MO. We're now scheduled to launch the GRS on a Mars Surveyor orbiter in 2001. The spacecraft will arrive at Mars in 2002, and will complete its job in 2004—nineteen years after I bought my nuclear physics textbook. You sometimes have to take the long view in this business.

By the summer of 1996, it looked like everything was pretty much back under control in the Mars business. The first Mars Surveyor spacecraft was on the way to the pad, the next couple were well into their design phase, and it seemed like we were about to settle down for a nice, calm, extended period of exploration.

And then, all hell broke loose.

In August of that year, I was vacationing with my family on the upper peninsula of Michigan. We had a little cottage on an island, and I spent a delightful week canoeing, building a tree house for my kids, and generally enjoying being away from telephones and computer networks for a while. Then, at our first overnight stop on the way back home, I figured I'd better see what had been going on in the world, and plugged in my laptop. My email program practically exploded. While I had been gone, scientists at Johnson Space Center had announced that they had found evidence of life in a martian meteorite.

I had known about this meteorite for a while. It's one of twelve that are believed to come from Mars. Certainly they all come from the same object, since details of their chemistry (for example, of the isotopes of oxygen) are all very much the same and are unique to this group of rocks. The young age of most of them hints that Mars was their source, since there simply aren't that many places in the solar system where young rocks can be found. But what seems to clinch the martian link is that one of them contains tiny bubbles of gas that is a near-perfect match to the composition of the martian atmosphere.

Of the twelve martian meteorites, one (with the romantic name ALH84001) is special. Alone among the set, it is old—nearly four billion years old. This makes it the only martian rock we have that comes from the early warm, wet period of the planet's history. With no other ancient rocks from Mars available to us, ALH84001 is the best place we now to look for evidence of ancient martian life.

Dave McKay and his colleagues at Johnson Space Center looked, and to their amazement they thought they found something. The rock was shot through with veins of limestone that suggested that water had once flowed through fractures in it. That was exciting, but hardly Earth-shattering. However, they also found that the limestone contained minute traces of organic molecules, hinting that the chemical building blocks of life had been present. Most intriguingly, the limestone veins contained tiny structures, far smaller than the tiniest microfossil ever found on Earth, that looked more than a little bit like fossilized bacteria. Or like fossilized *something*, anyway. None of these findings alone was enough to conclude that life had once existed on Mars, but finding them all together in one rock was enough to encourage McKay's team to make their announcement.

The announcement was an enormous media event, and it had a huge impact on the Mars program. The immediate result was that there were meetings—a *lot* of meetings. If nothing else, the announcement did wonders for the frequent flyer mileage of most of the Mars scientists in the United States. But out of those meetings came something real: a new direction and a distinct focus for the Mars exploration program. Before ALH84001, we had had a broad goal of understanding the planet in all its complexity. Our new mandate was simple: find out if Mars had once harbored life.

It's a shame that ALH84001 by itself probably can't answer the question. For one thing, even the McKay team admits that the evidence is inconclusive. As one of my colleagues noted, three or four maybes don't add up to a yes. More seriously, ALH84001 is *not* the rock that you'd choose to answer the question. The simple fact is that we don't have a clue where on Mars it came from. Old terrain, yes, but where? A lakebed? A valley floor? Or just some non-descript cratered terrain, far from anything that might have been a reasonable habitat for life? The answer matters, and we don't know it. ALH84001 was delivered to us by nothing other than good fortune. A meteorite impact somewhere on Mars knocked off a little chunk of the planet, and it happened to find its way to Earth and into our hands. But we didn't select the spot, let alone select the rock, and that's no way to search for fossils.

If you pick up a rock at random on Earth, the chance that it will contain a fossil is small. In fact, if you pick up a really ancient rock on Earth at random, the chances that it will contain a microfossil are almost zero. People spend careers searching for these things. But with ALH84001 we are faced with the possibility that the very first random ancient rock from Mars that we picked up happened to have microfossils in it. The idea stretches credibility, but at the same time the evidence is too provocative to ignore.

So ALH84001 is not enough. What is needed to answer the question is a set of rocks picked carefully and deliberately with the search for ancient life in mind. Maybe this means going to an ancient martian hotspring and looking for limestone deposits that could entomb microfossils. Maybe it means going to an ancient lakebed—the martian equivalent of Lake Hoare—and taking sediment cores there. But until we do it, and bring the samples back, I fear that the question will remain open. It'll take new roving vehicles that can traverse for long distances across the surface. It'll take new instruments and sampling tools to find and collect the right rocks. It'll take some pretty fancy new spacecraft to fly those rocks back to Earth. And, if the past is any guide, it'll take a lot of patience.

So, did life arise on Mars? I don't know, and I won't for a long time. But it's a profound question. If we find that it did, we will know that life on Earth is not unique in the universe, or even in our solar system. If we discover that it did not, despite the warm and wet conditions, we will have learned something very basic about the conditions required for life to begin.

And here's what intrigues me most: suppose we find that life did arise on Mars, somehow springing into being from inanimate material. How does such a miracle occur? It seems to have happened on Earth, but on Earth the record of genesis is gone, destroyed by our planet's relentless tectonic and volcanic activity. On Mars, though, almost half the planet dates from the first billion years of solar system history. Its early geologic record is still there to be read. Mars holds the promise then, faint but real, of being the one place we can go, at least in my lifetime, to learn how life begins.

Not a bad prospect, and one of the things that keeps me going back for more.

2

Venus: the way we might have been

ELLEN STOFAN

Ellen Stofan is a rising star. Within a few short years of completing graduate school, she had been both the Chief Scientist on NASA's New Millennium Program, the Deputy Project Scientist on the Magellan Project to Venus, and the Experiment Scientist on SIR-C, an instrument which provided radar images of the Earth on two shuttle flights. Before entering professional life, Ellen studied geology and art history at William and Mary University, and then earned Masters and PhD degrees from Brown University, studying the geology of Venus using Soviet spacecraft and Earth-based radar data. She has been awarded both NASA Exceptional Achievement and NASA's Exceptional Service Medals, and was identified as one of the "25 Rising Stars of Space" by the National Space Society; she was profiled in the 1994 IMAX film "The Discoverers." Ellen and her husband Tim Dunn have three children and are living in London, England. She provides us here with a truly unique look into the highs and lows of the Magellan Venus radar mapping mission, and along the way, explains just how little we seem to really understand about a planet we once thought was Earth's twin.

The first images of the surface of Venus were due around four in the morning on August 11, 1990, about a week after NASA's Magellan spacecraft went into orbit. Back on Earth, those of us who worked on the Magellan team were waiting at the Jet Propulsion Laboratory (JPL) in Pasadena, California. We were anxious to see if this mission, reshaped from an earlier more ambitious science mission and delayed for years after the Challenger disaster, was going to work. I was woken up at about 4:30 am by a phone call from JPL; it was Steve Saunders, the Magellan Project Scientist. His voice sounded dreadful. He said they were having trouble processing the images, and that the signal from the spacecraft had been lost. After a night spent dreaming of seeing aliens dressed in 1950s clothing in the first images (maybe we had too many tabloid covers taped up around the Magellan Science area!), I rushed into JPL feeling fairly desperate.

Why had this spacecraft been sent to Venus, and how had I ended up in a dark California parking lot nearly in tears over some hunk of metal millions of miles away? Venus is a planet that has fascinated people for centuries: the Evening Star. Named for the Goddess of Love and Beauty, its real nature is more malevolent, with an atmosphere of corrosive gases. The thick clouds of the venusian atmosphere have produced a runaway greenhouse effect, trapping the heat from the Sun and producing surface temperatures of up to 500 (!!) degrees Centigrade.

Venus has been called Earth's twin, and it truly is the most like Earth of any planet in our solar system. The two planets are about the same size and made of similar materials, as they formed relatively near to each other 4.5 billion years ago when all of the planets of the solar system formed. Venus and Earth differ in their surface temperatures, and the obvious resulting lack of water at the surface of Venus. These similarities and differences make Venus an excellent place to go to better understand our own planet. Why had two places that started out so similar ended up so different? To answer this question, we need to know what Venus is really like: what does its surface look like, what is going on in the interior of the planet? To study another planet, we need to examine its surface appearance and heights, measure the compositions of its surface and atmosphere, and gather geophysical data such as gravity measurements to peer into its interior. Much of our early knowledge of Venus came from the intense Soviet exploration program, which sent back images and rock compositions from the surface, and early US missions which provided information about the surface and atmosphere. But Magellan would be the first mission to gather a data set that would really allow us to study what the surface of Venus was like.

How does one end up wanting to study the planets? I had wanted to be a planetary geologist since I was about 14. I grew up outside of Cleveland, Ohio. My father was a NASA engineer, launching rockets, while my mother was an elementary school science teacher. When I was in elementary school, I wanted to grow up to be like the archeologist Mary Leakey, but became interested in geology after tagging along on my mother's graduate school geology course field trip. After hearing people like Gerry Soffen, Tim Mutch, and Carl Sagan speak at the launches of the Viking spacecraft, I learned that there was a way to combine my interest in the space program with geology (by the way, my sister rebelled against all this space stuff; she is a lawyer!).

School days

When I arrived at graduate school at Brown University in 1983 to work with Jim Head and Richard Grieve, I really had no focus, other than that I wanted to do "planetary." I sort of had a vague desire to study Mars (I had an undergraduate internship mapping a region on Mars), but I was open to anything.

I began with two projects—looking at terrestrial impact craters to support the flight of the Shuttle Imaging Radar-B (SIR-B) mission (a program that would come back to haunt me later, as I became involved in the SIR-C program at JPL) and looking at circular features in radar data of Venus collected at the Arecibo Radar Observatory in Puerto Rico by Don Campbell and others. SIR-B had a series of problems which led to our obtaining virtually no data, so I began to concentrate on Venus studies with Arecibo data.

At about this time, two Soviet spacecraft, Veneras 15 and 16, were sent on their way to Venus. These two spacecraft mapped the northern quarter of Venus in 1984 and 1985. Their mapping was limited by the spacecraft orbit; the radar images returned have a resolution of about 1 kilometer. Brown University had a program of cooperation with the Vernadsky Institute in Moscow, which was responsible for interpreting the data. All of this resulted in six exciting years exploring excellent, brand new images of Venus for my PhD thesis, with new discoveries around every corner.

A huge side benefit was the opportunity to collaborate with a number of Soviet scientists, including Sasha Basilevsky and Alex Pronin, and to observe the changes taking place in the then Soviet

Union between 1985 and 1989. Some of the geologists from the Vernadsky Institute spent much of their time ensuring that we got to see the richness of Russian art and architecture. My favorite excursion was to the village where Tsiolkovsky, the father of modern rocketry, had lived. The town is dedicated to space exploration. There's even a statue in the park of a heroic-looking woman holding a model of Sputnik! The space museum there was fascinating, especially seeing the cosmonauts' provisions for landings in Siberia, which included fishing gear in case it took too long for the ground crews to find them. My favorite aspect of the village, though, was the elaborately carved wooden houses, which reminded me of houses in the Baba Yaga fairy tales.

The scientists took us into their homes, and made us feel part of their families. I had to learn how to drink vodka "neat," and how to resist endless encouragement to empty a glassfull in one gulp (having seen the rapid effects of this on some of my colleagues, I resisted). For me, my five visits to the Soviet Union were an incredible time scientifically and personally, being able to utilize a newly collected data set, collaborate with fun and interesting people, and learn about a culture that, although my great-grandfather was Russian, I knew little about.

Of course, all of the Russian scientists spoke English, and only one of us knew any Russian. We did many of our formal sessions in Russian, having line-by-line translations done, which made it difficult to give a talk. I would say one line, then listen fascinated to the translator, then forget what I had just said. My favorite moment came when I went to eat what I thought was cheese on a cracker. After swallowing, I asked what it was. The person thought for a while, searching for the English word, and finally said "It is like bacon, without the red part." Lard cracker. Yum.

Magellan

When I finished my PhD in 1989, Steve Saunders offered me a postdoctoral position at the Jet Propulsion Laboratory in Pasadena, California. My husband was willing to quit his job and look for one in

California, so off we went with our 14-month-old son in tow. I arrived nine months before Magellan went into orbit around Venus. As I learned to become used to the relentless sunshine, smog and shaking ground of Los Angeles, I learned about the Magellan spacecraft and how it worked, and helped Steve prepare for the onslaught of scientists and data that were soon to come when Magellan reached orbit around Venus. I rapidly became incredibly impressed with everyone who worked on Magellan; they really exemplified the word "team." To this day, I have not worked on another project where there was so much respect for others, and so much of a genuine commitment towards a single goal: making the Magellan mission an incredible scientific success. The credit for this wonderful spirit was due to a number of people, particularly John Gerpheide and Tony Spear, the first two Project Managers of Magellan.

Back to that early morning dash into JPL. By the time I got there at a little after 5 am, the first images were processed, and the signal from the spacecraft was re-acquired later that day. The first strip of data that was processed was not particularly exciting, volcanic plains and half an impact crater, but for those who had worked so long and hard on this mission, it was like seeing the face of a new child. To look at the surface of another planet for the first time, seeing what no one has seen before, is an incredible feeling of awe and excitement.

John McPhee, who has written many excellent books about geology, spoke of geologists as living in "deep time," where we learn to regard millions of years as an instant. I think that planetary scientists have a different sense, one of "deep time and space," where not only our sense of time has broadened, but our sense of place has changed. We are not only of this Earth, we are of this solar system, of this universe. By visiting places with our spacecraft, our sensors, and our minds, we travel through time and space. We do not require the bravery of true explorers, just their curiosity.

Magellan's communications problem recurred, but was found to be caused by a software "bug" and was corrected. Over the next few years, Magellan worked nearly flawlessly, mapping over 98% of the surface of Venus. Much of our time was spent getting the first scientific results of the mission published in journals, getting the data set out to the whole scientific community, and communicating as much of our excitement as we could to the public.

To paraphrase a children's book, Magellan truly was "the spacecraft that could," built of spare parts and downsized, before the word became fashionable. The radar antenna, the main instrument on board, was a spare left over from the earlier Voyager mission. The radar instrument was crucial, as radar can penetrate through Venus' thick cloud cover to return images of the surface. Also on board was a radar altimeter, which measured surface heights, allowing a topographic map of the surface to be produced.

Magellan was initially funded for only one 243-day mapping cycle around the planet (which is one Venus day, or the amount of time it takes Venus to rotate once under the spacecraft). After that, it was a struggle to obtain funding for the spacecraft team for more mapping cycles. In these extended mission years, the spacecraft was used to fill in some unmapped regions and to look at the surface from different angles (important with radar, because the surface's appearance can change depending on view direction and angle). The extension of Magellan's mission also allowed us to perform aerobraking, where we used the atmosphere of the planet to circularize the orbit to obtain high-resolution gravity data. The investment in this exercise at Venus is paying off as this technique is now being used to reduce the fuel necessary to place Mars spacecraft in their proper orbits. After four years at Venus, Magellan came to a fiery end in the fall of 1994, as the orbit finally decayed to the point that the spacecraft entered the atmosphere and burnt up. By the time the mission ended, Magellan had collected more data than all previous planetary missions combined, and had returned a better basic global map of Venus (see Figure 2.1, color section) than we have for Earth (due to our inability to "see" the ocean floor).

What did we find at Earth's twin? Four years past the mission's end, most of us are still trying to sift through the data. First, let me discuss what scientists agree on. Then I'll discuss more controversial issues.

We see from the Magellan data that Venus has no system of plate tectonics as Earth has. Plate tectonics is the way we describe the motions of the outer layer, or lithosphere, of the Earth. The surface of Earth is broken up in to a number of large plates, which are basically in constant (but very slow) motion. Where plates come apart, as at mid-ocean ridges, new crust or surface is created.

To balance the creation of the new surface, plates are destroyed at subduction zones, like the Marianas trench near Japan, where they are pushed down into the interior of Earth. Mid-ocean ridges produce relatively quiet-natured volcanism, while subduction zones can produce very explosive volcanism, as at Mt. Pinatubo in the Philippines, or large earthquakes, as in Japan. The third type of plate boundary is where two plates slide past one another, as along the Pacific Plate–North American Plate boundary in California, better known as the San Andreas Fault. Obviously, slide is a bit of a euphemism; for those of us who lived through the Northridge earthquake of 1994, it felt more like two jagged plates jumping roughly next to each other!

Plate tectonics is related to the main way the Earth loses its heat. Heat is produced inside a planet by the decay of radioactive elements— the same process that produces radon gas in the basements of houses. All of this heat has to find its way out and, on a planet, heat loss is what drives the geologic processes that we find on the surface.

So how do we know that Venus has no plate tectonics? And what does this mean? On Earth, most geologic activity is concentrated at plate boundaries. For example, the edges of the Pacific Plate are called "the Ring of Fire," as they are delineated very well by volcanos. Maps of the distribution of volcanos, rifts, and mountain ranges on Venus show no organized pattern. This implies that there are no plate boundaries, and thus no system of plate tectonics on Venus. This in turn may be because Venus lacks water. Water lowers the melting point of rocks, "lubricating" the subduction process which really seems to drive plate tectonics on Earth. So, we have a fundamentally different planet than Earth. If Venus ever did have plate tectonics, no sign of it is left on the surface today.

We also learned from the Magellan and earlier Soviet and Earth-based radar that Venus had a lot of Earth-like geologic features. Venus is really a vulcanologist's delight. Very few areas on the planet have not been affected by volcanic activity. The low-lying plains of Venus are made up of long, broad volcanic flows, some of which extend for hundreds of kilometers. Volcanos with long, broad slopes, similar in size to Mauna Loa on Hawaii, are scattered around the planet (see Figure 2.2, color section). These larger volcanos are vastly outnumbered by small volcanic cones and shields, which sometimes occur in clusters of over fifty. "Small" is a relative term; many of these volcanos are 5–10 km across, and would cover nearly all of Manhattan!

Some of the volcanic features are quite bizarre. This sent many of us searching for easy words to describe them, which led frequently to some silly nicknames. Bizarre-shaped volcanic collapses were termed "gumboids" after the famous rubber figure, flat-topped volcanos were called pancakes, circular volcanos with leg-like protrusions were called "ticks," and volcanos surrounded by petal-like flows were called "anemones." These names usually start out as a kind of shorthand, after weeks of endless days of work, when your sense of humor is at strange levels. We have tried to purge most of these nicknames from the literature and describe features in a more clear and scientific (dry!) manner.

The strange appearance of some of these features may be due to the very high surface temperature and pressure on Venus. Our failure to predict some of these volcanic forms will ultimately help us to better understand the process of volcanism throughout the solar system.

Along with volcanic features, the surface has also been shaped by tectonic processes, which are horizontal and vertical motions of the surface. Long rift zones, where the crust of the planet pulls apart, cut across thousands of kilometers of Venus' surface. The whole equatorial zone of the planet is ringed with a series of rifts, called Dali, Diana, Hecate and Parga Chasmata. Some of these are quite similar to rift zones on Earth, such as the one in east Africa. Three large mountain ranges can be seen on Venus (Akna, Freyja and Maxwell Montes). These mountain ranges look folded, similar to patterns seen in radar images of mountain belts on Earth like the Appalachians. Smaller venusian mountain ranges, called ridge belts, are found in two low-lying plains areas. These smaller belts were formed by less intense compression of Venus' surface than that which gave rise to Akna, Freyja, and Maxwell Montes.

New types of geologic features, called coronae and tessera, were also identified. Coronae (Latin for crowns) are large circular features surrounded by a ring of fractures (see Figure 2.3, color section). The largest one, Artemis Corona, is over 1,000 km across. Coronae are thought to be produced when a plume or perhaps a blob of hot material rises from deep with the planet, bulging up the surface, similar to the process that forms the Hawaiian islands. I studied these features for my PhD thesis. Tesserae are high plateau-like regions cut by fractures and ridges trending in every direction. Their complex fracture patterns reminded the Russian scientists who first identified them of the parquet floors in their institute; this was later formalized into tesserae (tiles in Latin).

Akna, Diana, Artemis: where do these names come from? For all the planets, indeed all objects of the universe, strict nomenclature rules are followed. The International Astronomical Union's Working Group on Planetary System Nomenclature decided some time ago that all features on Venus should be named for women, quite properly in my opinion! The Working Group has set a series of rules: certain types of goddesses for certain types of features, famous women for impact craters, new feature types get formal Latin names. Coronae, for example, are named for fertility goddesses. (I have had to put up with numerous silly comments as I have spent most of my career studying coronae, and much of it pregnant with one or the other of my three children.) If the feature (an impact crater or volcanic caldera) is to be named for a famous woman, the individual must be dead for at least three years (therefore really truly dead), not a representative of any living religion, not a political or military figure of the nineteenth or twentieth centuries, and not a person of special national significance. In addition, the list of names should be representative of all cultures. It is nice to see women who have made great contributions to the arts and sciences receive acknowledgment, although they are obviously not around to know it! The lone male on Venus is James Clerk Maxwell, for whom the largest mountain range is named. The feature was named before the convention was set up (it was the brightest feature in radar, and hence was named for the man who had theoretically predicted electromagnetic waves).

Venus unveiled

The Magellan data that we have for Venus has left us with a mixed picture: Venus and Earth have great similarities and great differences. This is often the case in science; when things are not as you predicted, you start to learn how things really operate! One thing is clear, the clues to understanding the current state of Venus, and its implications for Earth, lie in understanding what has taken place on that planet for the last 4.5 billion years.

At the time of solar system formation, the planets accreted from primitive materials, those that make up asteroids and comets. As the rocky inner planets formed, they went though a process called differentiation, with heavier materials like metals moving towards the center of a planet to form its core, and lighter materials forming its outer layers. Water and other gases were released by heat from the interior during this process, forming the atmosphere of the planets. Venus' greenhouse atmosphere probably formed around this time, although isotopic ratios of gases in the venusian atmosphere may indicate that the very early Venus was a wet planet, perhaps even with its own oceans. However, we see no signs in the Magellan images that there were once oceans or rivers. Venus' greenhouse is mostly composed of carbon dioxide, not emitted from fuel-inefficient cars driven by early Venusians, but escaped from the interior of the planet during its early history mostly through volcanism (this is not to make light of the greenhouse effect on Earth: Venus should serve as a warning to us that greenhouse atmospheres do form, and do result in uninhabitable planets).

After that early phase, which we know only through theory, our knowledge of the next several billions years of the history of Venus is negligible. The average age of the surface of Venus is about 500 million years, not too different than the average age of the surface of Earth. This number is arrived at by counting the number of impact craters on the surface; older surfaces will have more craters. If we combine this with estimates of the number of impactors (asteroids, comets) that come through the solar system over time, we can use the impact crater population to give us an actual average age for a planetary surface.

Units with different numbers of impact craters were identified on the Moon, and assigned ages using this method. After samples brought back by Apollo were dated using laboratory techniques, more accurate ages could be assigned to some areas on the Moon. This allowed the whole technique of utilizing impact craters to date surfaces to be used more accurately all over the solar system, although you still have to make assumptions about impactor population variations in different regions of the solar system.

The craters on Venus tell us that the planet's surface has some mixture of ages, averaging out relatively young, at about 500 million years. Why is 500 million years young? This is where that "deep time" concept comes into play. All the planets formed about 4.5 billion years ago, so the surface of Venus is only revealing about the last 10% of the planet's history. The average age of the Moon's surface is about 3 billion years, for Mars the average is about 1.5 billion years; most geologic activity on these planets ceased at this time. So of the planets of the inner solar system, Venus, again, is the most like Earth. But, this is not surprising. Remember that Venus and Earth are very similar in composition and size, and thus should contain about the same amount of heat-producing elements. As it is internal heat that drives surface processes, Venus, like Earth, should be a geologically active planet.

So how can we put together a picture of Venus with its Earth-like geologic features, relatively young surface age, but without plate tectonics? This is not really an idle question, as we really do want to understand how one makes an Earth! We are starting to have the technology to detect planets beyond our own solar system, and hope within our lifetimes to have a picture of an Earth-like planet from around another star. But, is it hard to make an Earth? How could Venus and Earth have started so similar and ended up so different?

Right now, we have no answer to these questions. What has developed are two main theories on how Venus might have evolved, and a few ideas on the critical factors that might have made Earth a great planet to grow up on, and Venus an uninhabitable nightmare. The two theories can be labeled as "the catastrophic Venus" and "the Earth-like Venus."

The first theory has Venus shaped by a series of catastrophes, with the planet quietly awaiting its next upheaval. This theory, first advocated by Gerry Schaber and Bob Strom, arose from the initial analysis of the Venusian impact crater population, which indicated that the craters were statistically randomly distributed. This would suggest that all of the surface was about the same age. The only way that some scientists could come up with to explain this was that Venus had undergone some sort of major volcanic and tectonic upheaval about 500 million years ago, erasing all traces of its early history. Since then the planet has been relatively inactive, collecting impact craters. A newer modification to this theory, by Sasha Basilevsky and Jim Head, holds that the planet went through a series of epochs where a single geologic process dominated the planet: an "age of tessera formation" followed by a time when most large volcanos formed, and so forth. These ideas led to the development of some theoretical models which try to explain mathematically what is going on. These models include either an episode of very rapid plate tectonics using up lots of Venus' internal heat followed by a quiet period, or an interior of Venus that catastrophically overturns every billion years or so. This is quite different to how we think Earth operates. On Earth, geologic processes are not confined to any single time period, although they may vary in rate over time.

The other theory is that Venus behaves more like the Earth, with the same geologic processes operating over time. This theory is really the assumption that most scientists supported before Magellan, and that John Guest and I have been revisiting recently. A second look at the impact crater population by Steve Hauck, Roger Phillips, and Maribeth Price may indicate that the population is not really random. This means that no catastrophes are required. The volcanism and tectonism that wiped out the early history of Venus may have occurred at uneven rates over time, and this wiping out process may have happened over tens to hundreds of millions of years. Because of the lack of water on Venus, the surface expression of heat loss may be much more subdued than on Earth, producing more features like coronae and no real plates. I think this theory is stronger, as it seems to fit better what we see when we look in detail at the complex, interwoven history of volcanism and tectonics on the venusian surface.

The two theories remind me of debates held hundreds of years ago, when scholars tried to unravel Earth's history. Catastrophism, first based on biblical notions, held sway, with most geologic features thought to have formed in Noah's Flood. Features that looked the same (mountain ranges, outcrops of red rocks) were thought to have formed at the same time. Eventually, this was superseded by the "uniformitarian" school of thought that dominates Earth science today. Are we walking down the same path in our study of Venus, or is it the catastrophic planet we once imagined the Earth to be?

To really decide between these two theories, we need more time to analyze the wealth of Magellan data, and we need some very specific measurements made at Venus. We need to measure rock compositions at Venus, to better determine how the surface has evolved over time. We need to better understand the atmosphere: what is the greenhouse made of? How does it work? When did it form? We need to understand if Venus was ever wet, and what this might have meant for any kind of pre-biologic activity on the planet. We need to send seismometers to the surface to measure venusquakes, which will show us how the interior of the planet is layered. We need ultimately to bring samples back to Earth, or send miniature sophisticated laboratories to Venus in order to date the ages of surface rocks. Venus does require some special technology, though, as operating a lander at temperatures the same as those in a self-cleaning oven is no easy task!

On

As Magellan wound down, the project shrunk in size, to show that we could do things, not only cheaper and faster, but that all important "better," as well. I moved onto the SIR-C program as the Experiment Scientist, helping to turn data requests from over 30 scientists into a coherent science mission. It was great fun, but I never stopped missing the incredible experience of Magellan. In fact, after SIR-C, I became the Chief Scientist on the New Millennium Program, mainly because I want to see us going back to Venus soon, to answer some of these remaining questions. The New Millennium Program is devoted to infusing technology into NASA missions to lower their cost. Venus, with its extreme environment, will remain a difficult and costly place to do surface science, until we invest and capitalize on new technologies. It was this frustration that made me so enthusiastic about a technology-based rather than science-based spacecraft program.

My research still focuses around the basic questions about Venus. How does Venus lose its heat? How can we use its past history to understand how Earth-like planets evolve? How do its volcanos compare with volcanos here on Earth? These topics will keep me busy into the foreseeable future, and perhaps even occupy one of my children some day! (I think my kids love my job because they see me getting to go work in cool places like Hawaii and Mt. Etna, and they like to meet astronauts!).

My absolutely favorite task during the Magellan mission was interacting with the people who had paid for the mission, the public. Media interaction was always a kick. Highs for me were being interviewed on CNN and participating in a documentary on Venus. My media lows were being interviewed by a man in a clown suit for a local children's show (a good cause) and seeing myself in an IMAX-format movie—just think of yourself on a bad hair day 20 feet tall! My children, who got quite used to seeing me on TV, were a little alarmed to see Mom SO BIG!

I also spent a lot of time giving talks, sometimes as many as four per week at the height of the mission. From five year olds to senior citizens, I got questions that stumped me and ideas that made me question some of my own assumptions. Those talks always energized me, because the interest and support that I heard made me feel that Magellan had been a good investment. This was not only from a scientific point of view, but from letting everyone in on the excitement of being an explorer, of going beyond our world and finding the excitement of the process of discovery.

Through my time with NASA, I have occasionally been asked by those critical of the space program: "How will this really help us solve any of the problems we have here on Earth?". While some aspects of the space program have practical spin-offs, our exploration of planets is more open-ended in its benefits. Our studies of Venus may ultimately help us understand greenhouse atmospheres better and how they form, how volcanos operate, and how planets turn out like Earth. But, even more importantly, I feel that exploration gives us a sense of the wonder and vastness of the universe. It is an expression of one of the better aspects of human nature—our curiosity to explore and to know. One morning or evening, go out and look at that bright, bright star-like object hanging in the sky near the Sun. Imagine the swirling sulfuric clouds, and the searing heat of Venus. Enter deep time and space, and watch the formation of the twins of Venus and Earth, then see them travel down their divergent paths. You don't have to leave Earth: you are already a planetary explorer.

3

Moonlighting

CARLE PIETERS

Carlé Pieters. Ask a planetary scientist to name a few prominent lunar spectroscopists, and a name you're sure to hear is Carlé Pieters. Both Carlé's deep expertise, and her lab at Brown University in Providence, Rhode Island, are internationally recognized. Carlé's career began with a stint as a Peace Corps volunteer in Malaysia, after which she went to graduate school at MIT (Massachusetts Institute of Technology), and served as a researcher at the NASA Johnson Space Center in Houston, Texas. Since 1980 she has held a professorship at Brown University. Carlé has been deeply involved in scientific research related to asteroids, meteorites, and the Moon. Here, she gives provides a wonderful exposition on what being a lunar astronomer is like. And along the way, she shows us that old Luna still has many surprises left up her sleeve.

Observational astronomers, or most of them, hate the Moon. It is a demonstrable fact that astronomers model their working rhythms around the lunar monthly cycle. Scheduling of all the major telescopes in the world are based on whether observing nights occur in "dark" time or "bright" time. Bright time is the least desirable and is still night time, of course, but it is defined as when the Moon also graces the heavens, polluting the night sky with reflected sunlight. This scattered moonlight makes it difficult to detect faint sources. The coveted dark time is given to deep space galactic astronomers, while the dregs of bright time is offered to planetary astronomers, who are incorrectly perceived as somehow having stronger photons available since they come from closer sources. I am a lunar astronomer and love the bright night time.

In the course of my work, I've suffered through my share of frigid nights that permeate the fluffiest down jacket, cramped necks that protest unnatural positions, and disappointments from undesirable weather. Nevertheless, lunar astronomers are not often thought of as real astronomers. After all, anyone can see the Moon. Not only that, lunar features (dark maria and bright highlands) can be seen with the unaided eye! What, in heaven's name, could an astronomer possibly want to use a telescope to look at the Moon for? Especially when years and years ago so much expense was poured into the Apollo program resulting in 382 kg of lunar rocks returned for study in Earth-based laboratories! People have even walked on the Moon; why should we give it another thought when there are much bigger issues to study? Ahh, the abuse lunar astronomers often endure from our noble but misguided colleagues who study objects much deeper in space. Lunar astronomers are a dying, or more accurately a transmorphing, breed. I'll explain in a moment.

Let me first tell you a thing or two about the Moon, our Moon. Everyone's most loved planet is naturally the Earth. The origin and continuation of our very existence depends on the temperate and water-rich environment maintained at our special location in the solar system, at 1 AU to be specific. [The unit of distance used by astronomers, an astronomical unit (AU), is defined to be the distance from the Earth to the Sun.] This particular part of the solar system is special in the Goldilocks sense—not too hot and not too cold, not too hard and not too soft. We recognize the Moon, as all our ancestors have, as a constant and familiar companion. Only in the last few decades, however, have planetary scientists come to appreciate how intimately intertwined the origin and destiny of the Earth and the Moon are. Random incidents at the dawn of solar system formation set in motion a series of irreversible events of enormous consequences. These ultimately led Earth's nearest planetary neighbors, Venus and Mars, to lose whatever water they had, and become uninhabitable, while Earth enjoyed life-harboring oceans and stability. Geologically active Venus (Earth's "twin" in physical properties) became choked with a heavy carbon dioxide atmosphere 100 times as dense as Earth's and little Mars became a beautiful but barren desert.

The early solar system was a very violent place. Scientists are still trying to discover what happened to Venus and Mars by extrapolating from what we think happened at 1 AU where the Earth and Moon currently reside. After the Sun formed in those early times, the proto-

Earth was growing by accreting solid material in this part of the solar system, as were several smaller proto-planets. A similar but smaller body (some estimate almost Mars-sized) chanced to collide with the proto-Earth and this interaction released enormous amounts of energy. Some of Earth's early mantle was ripped off and probably vaporized into the near-Earth environment, as did much of the stray impactor itself. The Moon is believed to have formed in Earth orbit from the debris in the aftermath of this cosmic cataclysm. Although lunar rocks are thus clearly reprocessed material, they still contain the tell-tale signature of their genetic relation to Earth (found largely in isotopic compositions). By the time Earth's oceans appeared, the Earth/Moon system had became stable and the influence of these two worlds on each other was permanent.

Farside

In spite of their common origin, the two planetary bodies of the Earth–Moon pair evolved quite differently during the first billion years. Formation of a primary solid crust is poorly understood in planetary evolution and probably depends on the specific circumstances of the planet, but once the crust is formed, the planet begins to take on character. After the initial period of violence, the Earth emerged with oceans while the Moon, large enough to be a planet itself, had a rocky crust stripped of its volatiles.

We actually know very little of what occurred on early Earth because the record of Earth's early crust is gone, overwritten by later events. Most of what we know about the earliest events comes from the well-preserved record on the Moon. The heaviest violence in the early solar system, or "heavy bombardment," ended about 4 billion years ago with a dramatic decrease of objects of different sizes pelting each other. The cratering record on the Moon provides an important yardstick to understand the timing of events for the rest of the solar system. Some of the largest scars on the Moon, numerous basins 500 to 2,500 km in diameter, are estimated to have been formed 3.85 to 4.2 billion years ago. Similar events must have also occurred on Earth in the first 500 million years but, as I've mentioned, the record has been erased.

When much of the material of the solar system had been either incorporated into planets or cleaned out, internal evolution of the planets began. Although only one-quarter the diameter of Earth, the Moon's share of inherited radiogenic elements provided sufficient internal heat to melt the lunar interior. Exactly how much melted and how deep or thoroughly is not yet known. What is known is that lavas of partial melt flowed from the lunar interior and filled many of the lowest areas on the Moon with dark basalt containing more iron-bearing minerals than the primitive lunar crust. Most of the holes left by the big basins on the nearside of the Moon were filled with basalt during this early period of profuse volcanism. The contrast between the bright highland crust and the smooth and dark basaltic mare (or "seas") creates the calm and friendly face we now see from Earth when looking at the full Moon. (Some see a sitting woman or a rabbit, but my brain has never made that association with these markings.) Almost all lunar volcanism ended about 3 billion years ago, leaving large areas of a pristine surface to record the subsequent history of our part of the solar system and to collect remnants of earlier debris and comets passing through.

Tidal interactions locked the Moon's rotation to the period of its revolution around the Earth so that the same side of the Moon always faces Earth. Just when this occurred is also not really known, but it was probably during those formulative first 500 million years, 4 billion years ago. Before going on, however, let me correct one enormous misconception that has somehow become embedded in our vocabulary: there is no "dark side" of the Moon. All sides of the Moon receive the same amount of sunlight. The lunar days and nights are relatively long; the Sun is above the horizon for 14 Earth-days and below the horizon for 14 more. (Since there is no atmosphere to scatter light, lunar night and lunar day would both be quite attractive for any lunar-based astronomer.) Because of this synchronous rotation, there is however a "nearside" of the Moon, which we always see, and a "farside," which we never see except by spacecraft.

The farside of the Moon is mysterious and poorly known. A few spacecraft circled the Moon returning images a few decades ago, preceding and during Apollo, but scientists have not been able to look at it with modern instruments until recently. A small spacecraft, called Clementine, was sent to the Moon in 1994 as a low-cost technology demonstration mission. Clementine carried an altimeter which showed that the farside has both the lowest and the highest elevations on the Moon. It turns out that these two extremes are associated with the largest recorded basin in the entire solar system, the 2,500 km diameter South Pole–Aitken Basin on the lunar farside. This is not actually a real basin name, but only a designation because the basin reaches from the South Pole to the crater Aitken, 15 degrees from the equator. For some completely unknown reason, this huge ancient hole was never flooded by the lavas that filled most of the basins on the nearside of the Moon. The farside thus looks quite different from the familiar dark-light markings we always see on the nearside. If you were a traveler in a spacecraft approaching the Earth/ Moon system with the Sun behind you and the Moon between you and the Earth, you would see the farside as a thoroughly cratered surface with a gray circular smudge between the equator and the south pole associated with this gigantic basin. The farside would look something like the Clementine image on the right in Figure 3.1. The reason this incomparable big depression is somewhat darker is an area of active research (its heavily cratered surface is not smooth enough to be mare). The moderately dark interior of South Pole–Aitken is thought to be due to deeper, more iron-rich minerals of the primitive lower crust exposed when the basin was formed around 4.2 billion years ago. We'll discuss this further later.

Lovely Luna

Let's move away from the early history of the Earth/Moon for a bit. Why is the Moon an interesting, and even fascinating, planetary object? I got hooked on the Moon when, as a graduate student wandering the halls of MIT, no one could answer a simple question about what caused some features I'd seen in a photo just returned from Apollo (dark smudges draping across a lunar mountain). One question led to another, and I soon learned there are far more questions than there are answers.

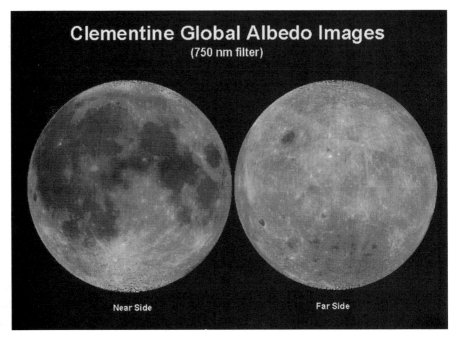

Figure 3.1

Albedo map of the Moon produced from digital Clementine images acquired over a period of two months in 1994. These data are mosaicked into a Lambert (equal area) projection (image made available by the Lunar and Planetary Institute at: http://cass.jsc.nasa.gov/research/clemen/clemen.html). The nearside hemisphere is the part of the Moon seen from Earth. Many of the circular basins on the nearside are filled with basaltic lavas forming dark maria. In contrast, the farside of the Moon contains very few maria. South Pole–Aitken Basin dominates the southern farside, but is only evident in this image as a roughly circular area of slightly lower albedo.

Three things have irreversibly held my continued enthusiasm and fascination with the Moon. The first, as I've already touched on, is the intimate and interwoven connection between the Earth and the Moon. The Moon is the only link to Earth's early history as well as the evolving environment in our habitable part of the solar system. The second is the recognition that the Moon, by its very proximity to the Earth, has been and will always be our springboard to the universe. Lunar exploration was started by the US and the former Soviet Union; the next wave will be led by numerous other players from the now international space club. At some point, interest in the Moon will inevitably evolve

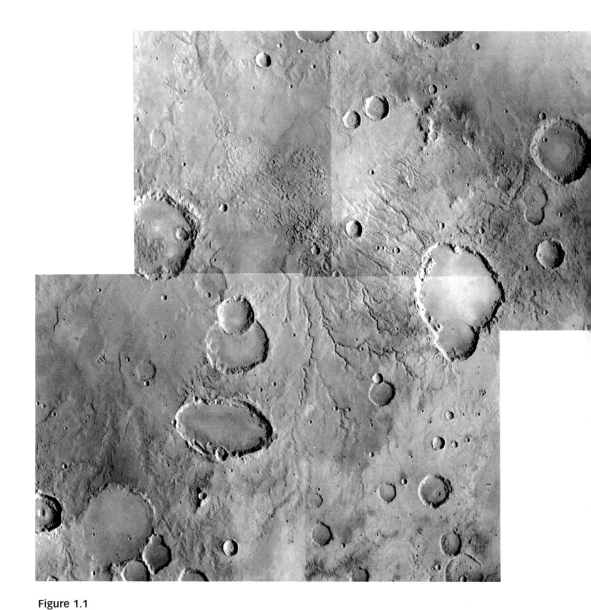

Figure 1.1

An ancient valley system on Mars.
Photograph from NASA.

Figure 1.2

SCUBA-diving in the dry valleys of Antarctica.
Photograph by Dale Anderson.

Figure 1.3

The ill-fated
Mars Observer
spacecraft.

Figure 1.4

Mars Pathfinder panorama of the martian surface. Photograph from NASA.

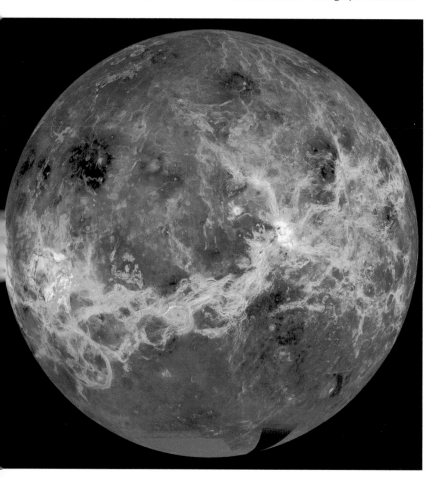

Figure 2.1

Magellan radar images were used to compile this global view of Venus, centered at 270 degrees east longitude. The simulated orange hue is based on color images obtained by the Soviet Venera 13 and 14 landers. Most of the surface of Venus is made up of low-lying volcanic plains. Across the equatorial region, a series of bright linear features are seen. These are rift zones, where the planet's crust is being pulled apart. Magellan image P-39225.

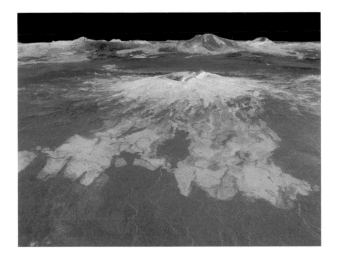

Figure 2.2

This computer-generated perspective view of the volcano Sapas Mons shows bright, rough lava flows extending from the summit of the volcano. Sapas Mons is about 400 kilometers across and stands about 1.5 kilometers above the surrrounding plains. The volcano Matt Mons can be seen in the background. The orange color is based on images from the Venera 13 and 14 landers. The vertical scale in this image has been exaggerated 10 times, in order to show the large, low-relief features. Magellan image P-40176.

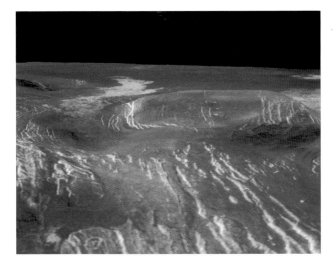

Figure 2.3

This corona, called Idem-Kuva, has a central bulge surrounded by concentric fractures, characteristic of most coronae. Bright, rough lava flows extend to the north from Idem-Kuva. This feature is thought to form over a hot mantle plume or blob that has risen to push up the surface and produce lava flows. The orange color is based on images from the Venera 13 and 14 landers. The vertical scale in this image has been exaggerated about 20 times, in order to show the large, low-relief features. Magellan image P-38723.

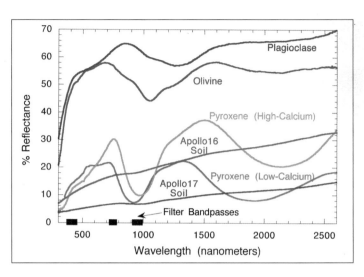

Figure 3.2 (above)

Visible wavelength (300 to 700 nm; nanometers) to near-infrared (700 to 2,600 nm) reflectance spectra for lunar materials and soils. The Moon reflects solar radiation through this part of the spectrum and diagnostic absorption properties of constituent minerals are incorporated in the reflected radiation. Most of the absorption bands are due to Fe^{+2} in specific crystal structures. Three of the five filter bandpasses (415 nm, 750 nm, 950 nm) for the Clementine UVVIS camera are shown at the bottom of the figure.

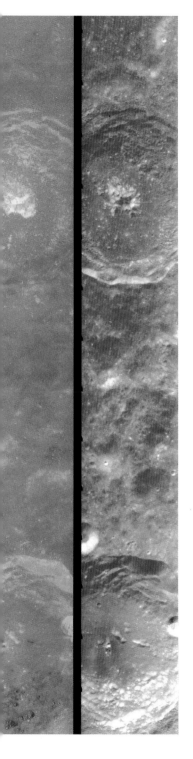

Figure 3.3 (left)

Multispectral images for an area through the center of the South Pole–Aitken basin located on the lunar farside.
Right: Calibrated brightness image at a wavelength of 750 nm.
Left: Color composite image created to distinguish between different soil and rock types. This image is formed from blue = 415/750 nm color ratio, green = 950/750 nm ratio and red = 750/415 nm ratio. The inverse relation between the red and blue channel causes areas to appear relatively red or blue according to the overall continuum slope of surface material. Any green-yellow-turquoise tones in the image indicate the presence of abundant iron-bearing minerals such as pyroxene. Bright areas appearing deep purple near the bottom of the image indicate almost pure plagioclase minerology.

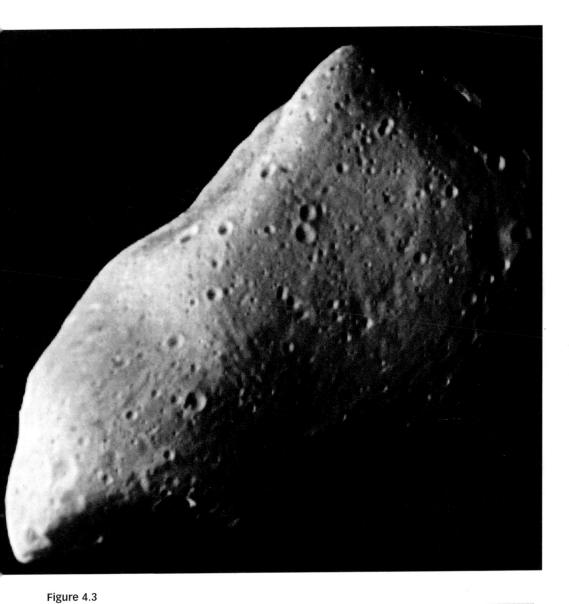

Figure 4.3

Highest resolution portrait of 951 Gaspra obtained by the Galileo spacecraft, showing its angular shape and the dearth of large craters. Photograph from NASA.

Figure 4.4

False color image of Ida and its moonlet Dactyl, constructed from Galileo images. Small, recent craters and patches of ejecta from a large, fresh crater on the backside of Ida look bluer than the average color of Ida, indicating that a "space weathering" process is modifying Ida's colors. Photograph from NASA.

from a scientific focus to a long-term vision centered on resources and utilization. And the third more personal stimulus is the glorious purity of the Moon's landforms. Anyone who knows and loves the desert will understand this intellectual and perceptual delight. The Moon probably won't excite the weather forecaster, or the biologist, but it's a geologist's heaven: a pure story of planetary formation and evolution written in the rocks. All the physical processes affecting a planet are laid bare with nothing to disturb it further. Its riddles are there to be read. (It turns out there is also soil, or a fine-grained regolith, formed from local rocks by processes active only in the space environment. And the distinctive lunar soil, with no Earth counterparts, has many stories to tell as well.)

A record in the rocks

So, this leads us from astronomy to geology. Rocks are, quite literally, the building blocks of solid planets. With only a few exceptions, the minerals that make up the rocks on the Moon are well known to Earth-based scientists. A mineral's composition and structure (how different kinds of atoms are arranged and bonded to each other) gives it its own identity. Nature doesn't care whether minerals are formed on Earth, the Moon, or elsewhere in the solar system; a given mineral is always put together the same way. Minerals such as plagioclase feldspar ($CaAl_2Si_2O_8$, with various amounts of Na substituting for Ca) and pyroxene [$(Mg, Fe)_2Si_2O_6$, with Ca regularly substituting for Mg or Fe] have the same fundamental and easily recognized properties regardless of their place of origin. It is the combination of such silicate minerals, their accessories, and trace amounts of minor elements that are different planet to planet and tell much about the formation conditions and processes involved in planetary evolution.

Silicates are by far the dominant minerals comprising both the crusts of planets as well as the stony meteorites which fall to Earth from elsewhere in the solar system. The structure of each mineral is one of many specific arrangements of bonds between the basic silica tetrahedra and other atoms which provide charge balance. A single silica tetrahedron is one silica ion (Si^{+4}, under normal conditions with a +4 charge) surrounded and bonded symmetrically in three dimensions by negatively charged ions that form a kind of cage around it. A happy arrangement is four oxygen ions which fit snugly around the silica ion forming a tetrahedron (a pyramid with the silica in the middle and oxygen at the four corners). Oxygen, however, normally attracts two electrons to be stable, giving it a -2 charge (O^{-2}). The complexity of rock-forming silicate minerals comes when other ions (Fe^{+2}, Mg^{+2}, Ca^{+2}, and Al^{+3}) are brought into the combination and different structures are formed with independent silica tetrahedra or chains of tetrahedra. The specific combination depends of course on the atoms available, the size of their ions, and the general properties of the environment. The formation of and interrelation between rock-forming minerals tell much about the pressure, temperature, and timing of geologic events. Reading the mineralogy of the Moon tells a fascinating story of what happened during those earliest times.

The ancient highlands of the Moon, the primary crust which encompasses most (80%) of the current surface, is believed to be dominated by the mineral plagioclase mentioned above. This calcium-aluminum-rich mineral is not very absorbing of visible light; hence, the lunar highlands are relatively bright. Current estimates based on lunar samples and limited remote sensing indicate the highland crust is composed of 70% plagioclase feldspar with the remainder often containing pyroxene of the low-calcium type. Now, an individual rock composed of 70% plagioclase feldspar is not that unusual; on the Earth such rocks are highly evolved and require special conditions for formation. But to have almost the entire surface of a planetary body with that composition is extraordinary, and demands a cosmic explanation. This mineralogy is a major clue to the formation and evolution of the lunar crust.

All lunar highland samples returned to Earth-based laboratories are breccias, ancient rocks battered and broken during the early heavy bombardment. A well-reasoned story of their original formation and evolution has nevertheless been developed by geoscience detectives working with small amounts of returned lunar samples. As Earth was readjusting from the original disruptive event, the scattered material accreted to form the Moon into a planet-sized body of 1,700 km radius.

The earliest property of the young Moon that can be surmised from the record is believed to be an ocean of magma, or melted rock, that extended down at least the outer few hundreds of kilometres. It is not known whether the heat source that produced this "magma ocean" was the energy released during accretion of the Moon or some other form of energy, such as from the decay of short-lived radioactive elements or from a massive outburst of the young Sun. The primitive lunar crust and the mantle that resulted as this ocean of hot rock cooled nevertheless carries its geochemical signature through the later period of heavy bombardment to the present rocks that now reside in Earth-based laboratories.

The magma ocean explanation for the plagioclase primordial crust also encompasses an explanation for properties of the lunar mantle, which, after all, is the source region for the dark basalts that filled the nearside basins. Although it has taken years of meticulous analyses to perfect, the magma ocean hypothesis for formation of the lunar crust and mantle is simple to describe. The early Moon had lost most of its volatiles (an observation that still remains unexplained), and minerals formed from the cooling magma ocean under reducing conditions (that is, minerals combined from available elements with as few oxygen ions as possible; leaving no water and no heavily oxidized rust). As the magma ocean cooled, at first the more dense minerals, olivine [(Mg, Fe)$_2$SiO$_4$] then pyroxene, formed and settled downward to create an iron- and magnesium-rich lunar mantle. The composition of the magma ocean thus evolved with time as minerals were formed and removed. When the melt reached the right composition, calcium-rich plagioclase began to crystallize and, because this mineral is less dense than the melt, it began to float and accumulated at the top. This process continued until a primitive feldspathic (i.e., plagioclase-rich) crust was formed that gradually thickened and became stronger. The lunar mantle below the crust consisted of minerals rich in magnesium and iron. Current estimates of the thickness of the lunar crust are about 60 km on the nearside of the Moon and thicker but averaging about 70 km on the farside (although South Pole–Aitken creates an anomaly).

The Moon has retained this primary crust through billions of years of subsequent history of the Earth/Moon system. The early crust of the Earth, on the other hand, has been lost through several periods of recycling through plate tectonics, and the crust we now live on formed much later than all the rocks of the Moon. (The geologically active process of plate tectonics, incidentally, has not been observed anywhere else in the solar system, which adds another component of uniqueness to the Earth/Moon system.)

Solidification of the lunar magma ocean was complete within a few hundred million years of the formation of the Earth/Moon system. By at least 4.2 billion years ago the lunar crust was rigid enough to retain the scars of the later period of the heavy bombardment that was apparently pervasive throughout the solar system. The dark iron-rich lavas that filled the lowlands of the Moon are the products of remelting of the lunar mantle largely between 3 and 4 billion years ago. The basaltic lavas on the surface thus reflect some of the properties of the interior lunar mantle.

Take note of all these various numbers and dimensions. If you try to put them all together you'll see why the Moon holds such a premier status as a natural laboratory for understanding how planets work and how the solar system evolved.

Since you've so patiently waded through part of the story of lunar mineralogy, I want to indulge in sharing some of the current excitement derived from the little bit of other new data acquired by a camera on the Clementine spacecraft, like the whole-Moon composite image in Figure 3.1. To do so, we must seriously discuss the color of the Moon. In addition to being an astronomer and geologist, I am also a spectroscopist. I pull apart the colors of Moonlight. I do this as a NASA supported PI of lunar mineralogy. (NASA research at universities is carried out through an annual process of peer-reviewed proposals made by individual Principal Investigators, or PI's.) I have studied the color of the Moon using telescopes for decades.

The color of the Moon

Let's investigate the topic of spectroscopy, and some of the useful information it contains. Although the human eye–brain system is an extraordinary analogue (i.e., nondigital) device, it is limited in that it sees only a small part of solar radiation reflected by the Moon. The Moon looks bright and almost colorless to us. But both impressions are wrong. The Moon is bright relative to the stars (it's closer, of course), but lunar soils in the laboratory are relatively dark compared with many terrestrial materials. Lunar soils are dark relative to ground-up lunar rocks as well! Neither is the Moon colorless, or gray, over the part of the spectrum that the human eye detects, from blue light at 400 nm (nanometers) to red light at 700 nm. In reality the Moon reflects an increasing amount of sunlight from the blue to the red, making it technically a "red" object, but this increase is not readily detected by our eyes. Using electronic detectors that are sensitive from the visible to longer wavelengths, we learn that the lunar reflectance continues to increase into the infrared. If the entire spectrum of reflected radiation is considered, the Moon is one of the overall reddest objects in the inner solar system (even redder than Mars in the near-infrared, which just appears red to the human eye because Mars has rusty minerals which reflect very little blue radiation in the visible part of the spectrum).

Now, spectroscopy of lunar materials has been studied in the laboratory with lunar samples, at the telescope with cameras and spectrometers, and just recently with spacecraft equipped with modest spectral sensors. A key result is that, superimposed on the characteristic red lunar continuum beyond the visible in the near-infrared, are several distinctive regions, broad bands of the spectrum, where light is absorbed by mineral components of the surface. These "absorption bands" are caused by specific ions, such as Fe^{+2}, in an ordered crystal environment and are highly diagnostic of the minerals present. The very regular structure of crystalline minerals makes their spectral properties very predictable. Absorption bands measured remotely thus have a direct compositional interpretation.

Two forms of spectroscopic data have been useful for lunar astronomers, and now, with the advent of advanced sensors on spacecraft, for lunar scientists in general. The first, multispectral imaging, or imaging at different wavelengths, has provided key information about diversity across the lunar surface. In this approach, digital images are acquired using several filters carefully chosen to be sensitive to specific subtle color variations in lunar materials. Each filter allows light to be measured over only one small part of the spectrum, or bandpass. Images taken at several different bandpasses can be calibrated and processed digitally in a computer. If one such digital image is divided by another, the resulting ratio image can map the small but important differences in color across the area imaged. The second approach evaluates the full spectroscopic character of material through continuous spectra of reflected sunlight. Good spectra consist of reflectance measurements at hundreds of closely spaced wavelengths designed to detect and characterize with high precision any absorption bands that may be present. I'll give examples of each.

Multispectral imaging of the Moon has been successfully carried out with digital cameras both by astronomers and by spacecraft. Measurements of lunar reflectance spectra, typically requiring more sophisticated optics, have been acquired in the laboratory using returned lunar samples and at the telescope using an evolving system of capable spectrometers. Although several such devices have been designed and selected for flight, unfortunately none of the advanced spectrometers has yet been placed on current spacecraft that have been or will soon be sent into lunar orbit (Clementine, Prospector, and the Japanese Lunar-A).

The three minerals already mentioned (plagioclase, pyroxene, and olivine), plus a few iron-rich and titanium-rich opaque minerals, dominate the mineralogy of most rock types on the Moon. It is the presence and relative abundance of these minerals and their location on the Moon (geologic context) that tell the story of lunar evolution.

Laboratory spectra of these minerals and two representative lunar soils are shown in Figure 3.2 (see color section) along with three of the five bandpasses used by the Clementine UVVIS digital camera at 415, 750 and 950 nm. As I mentioned earlier, the visible wavelengths detectable by the human eye run from 400 nm to about 700 nm. Diagnostic absorption bands of minerals occur beyond these visible wavelengths but are present in the near-infrared. Most of the near-infrared absorption bands seen in Figure 3.2 are due to Fe^{+2} in the structure of these various minerals.

The two broad absorptions of pyroxene near 1,000 nm and 2,000 nm tend to dominate lunar spectra. Both absorptions occur at shorter wavelengths for the low-calcium variety of pyroxene found in highland breccias and at longer wavelengths for the high-calcium variety found in basaltic material of the dark lunar maria. Two additional Clementine filters, at 900 nm and 1,000 nm, were included with the UVVIS camera to try to detect the shape of the prominent absorptions near 1,000 nm (the sensitivity of the detector limits this camera to 400 to 1,000 nm). The structure of olivine, a primary mineral of the lunar mantle, causes a broad multiple absorption centered at longer wavelengths than that of pyroxene (the absorption is also typically weaker). Plagioclase, the principal mineral of the lunar highlands, is the most reflective at all wavelengths. Although it is a calcium-aluminum silicate, the structure of plagioclase can accommodate tiny amounts of iron. The example shown in Figure 3.2 contains approximately 0.3% FeO, enough to produce a minor absorption near 1,300 nm.

Although the five filters of the Clementine UVVIS camera of course do not have the spectral resolution or coverage to characterize the mineralogy of the Moon, broadband imaging is often used to obtain some first-order information about the general properties of surface material. This information can be parameterized and displayed in a color composite, or "false color" image for visualization. Figure 3.3 (see color section) is such a color composite image; it displays real color variations, but not in a "natural" display. For example, if a ratio image is produced from images taken with the 750 nm and 950 nm filters, it provides information about the relative strength of the Fe^{+2} absorption in pyroxenes and olivines, and hence abundance. To a first order, an image of this ratio maps the distribution of these iron-bearing minerals. Albedo, or reflectivity, of lunar materials provides a second important piece of information. Plagioclase-rich materials are relatively reflective. The brightness of plagioclase-rich materials is true even for the well-developed lunar soils which themselves are darker than their pure mineral constituents (due to "space weathering'); the feldspathic highland soils are substantially brighter than the basaltic mare soils. A third parameter, the overall slope of the spectrum (or continuum), is sensitive to both composition as well as soil development and can be estimated by a ratio of the 415 nm and 750 nm reflectance. Mature well-developed soils are relatively red (high 750/415 nm ratio), whereas fresh feldspathic debris surrounding a recent crater has a less steep continuum over this wavelength range.

The color composite Clementine mosaic produced from these three spectral parameters using digital images obtained at 415, 750, and 950 nm is shown for a region within the farside South Pole–Aitken basin in Figure 3.3 (color). Each spectral parameter is assigned to a separate color (red, green, blue) for display. Even a glance at this image tells us that there is more to the story told by lunar color than meets the unaided eye when viewing the Moon.

One of the compelling lunar science questions asked by anyone familiar with South Pole–Aitken basin is: What part of the lunar interior has been exposed in this 2,500 km scar left by this major impact? In particular, did the impact event which formed the huge basin penetrate through the crust and excavate the ancient lunar mantle? On the Earth, geoscientists drill deep into the crust to understand the stratigraphy that has developed at a given site. The mantle is not readily accessible by drilling. In places, plate tectonics appears to have emplaced older fragments of lower crust or upper mantle in more accessible locations. On the Moon, our glimpse into the interior is to make use of the natural bore holes made by impact craters of various sizes. We know from experience and observation that larger craters excavate deeper. By examining the interior and deposits of craters of various sizes, we have (randomly spaced) windows into the interior.

Whole Moon

So, what does the largest crater in the solar system, the farside South Pole–Aitken basin, tell us about the makeup of the outer portion of the Moon?

Let's examine Figure 3.3 (color) more closely. Since the 750/950 nm image is displayed in the green channel, areas that appear yellow-green-turquoise have a high abundance of iron-bearing minerals. Throughout the area one notices a yellow-turquoise tone at fresh craters and other surfaces that have not developed mature soils due to steep topography. Separate analysis of data for all five filters indicates that the iron-bearing mineral is abundant low-calcium pyroxene instead of olivine or high-calcium pyroxene (based largely on the wavelength of the absorption band near 1,000 nm). This same mineralogy appears at both large and small craters, indicating no significant compositional variation with depth. Mountains found in the center of large craters have been brought from the greatest depth. For the large 87 km crater at the top of the image these deep-seated materials are rich with pyroxene and relatively homogeneous, quite similar to several of the surrounding features. On the other hand, the large crater to the south near the bottom of the image has several small pyroxene-bearing central peaks, but one which is different. The linear shaped peak on the right (easily seen in the gray tone image) is bright, but exhibits no notable absorption from iron-bearing minerals (deep blue in the color composite) indicating an almost pure plagioclase composition. A similar plagioclase-rich composition is seen throughout the south rim of this crater.

What is surprising about this mineralogy of South Pole–Aitken basin is the lack of evidence in this area for minerals suggestive of the lunar mantle—olivines and high-calcium pyroxene. The area is clearly rich with iron-bearing minerals, but the mineral assemblage is believed to be associated with the lower lunar crust rather than the lunar mantle. The remnant fragments of pure plagioclase within South Pole–Aitken is also surprising since most of the plagioclase-rich upper crust was expected to have been removed. Although South Pole–Aitken will inevitably be studied more thoroughly using additional Clementine data, the tentative answer to one question suggests a surprisingly shallow excavation by this enormous basin-forming event. If that conclusion holds, it means we have to ask additional questions about how impact basins are formed, how material is distributed in an impact event, and how a body responds after such an event. Even this little bit of new data from Clementine has caused scientists to reevaluate our meager knowledge about the Moon and the processes active on it and the other planets.

Working in Moonlight

Well, by now you probably have surmised why there may be fewer and fewer professional lunar astronomers. Not that there aren't important questions to address; quite the contrary. It is simply that access to new scientific data is tending to shift from that acquired by telescopes to that acquired by spacecraft. Although calibration of the Clementine data is not perfect, these digital image data are now global, an order of magnitude better in spatial detail, and waiting for analysis on a set of CD-ROMs. Expanding the horizon further with equally tempting possibilities are ground-breaking data designed to be acquired by small missions sent to probe unexplored corners of the solar system, including other enigmatic moons and planet-satellite systems.

The little bit of new data acquired by the first satellites to return to the Moon in decades has taught us (again) how little we really understand Earth's companion and our place in the solar system. South Pole–Aitken basin doesn't even exist on most maps of the Moon! (By the time new maps are made, perhaps the basin will be given a real name.)

When data and information are limited, of course so too is knowledge. With the new pulse of data from Clementine, we must reassess the boundaries of our ignorance. As peppered throughout my lunar rambling here, there are innumerable large gaps in our understanding of how the Moon and the Earth/Moon system formed and evolved. And these are not trivial questions. Planetary scientists understand the importance of this natural laboratory for understanding our home at 1 AU. Engineers and long-term planners recognize the Moon's value as a springboard to the permanent presence in space.

Lunar astronomers may be gradually transmorphing into planetary scientists dependent on data from space missions, but the process is nowhere complete! The amount of lunar data available from spacecraft is scanty and there is much to do; astronomy will continue to play a natural role leading science of the Moon and eventually on the Moon. On the practical or concrete side, we have a compelling need for high spectral resolution spectra that can address questions of lunar mineralogy more explicitly. Lunar resources are only vaguely known and spacecraft have not yet capitalized on current technology to meet this need. There are instruments to build and test at the telescope. A high-precision imaging spectrometer (or "hyperspectral" imaging system with the hundreds of near-infrared spectral channels needed for spectroscopy) has yet to target the Moon. There are different parts of the electromagnetic spectrum to explore. The flexibility of designing and redesigning an experiment using an Earth-based telescope cannot be supplanted by a fixed sensor on a spacecraft.

And there is just the simple joy of looking though a telescope at familiar terrain with ever-changing shadows. Photons are constantly streaming down to us from the Moon. They give us quiet pleasure on a moonlit night—and they continually test our ingenuity and creativity in devising ways to extract information from their energy.

Yes (among other things now), I am at heart a lunar astronomer, and I will always love the bright-time night time.

4

Small worlds, up close

CLARK R. CHAPMAN

Clark Chapman. Deep thinker. Old hand. Charmer. Wizard. These are the words that come to mind for Clark Chapman. Born and reared in New York (as are a surprising number of the authors in this book), Clark was educated at Harvard and MIT. After completing his PhD, he moved to Tucson and dedicated his career to the study of the belt of debris collectively called the asteroids which orbits in the emptiness between Mars and Jupiter. A scientific life dedicated to the study of asteroids might seem very narrowly focused but, as we learn below, beyond being endlessly fascinating in their own right, asteroids retain important clues to the birth of the solar system, to impact catastrophes on the Earth, and to the way in which objects "weather" in space. In 1996, Clark and his wife, LYnda, moved to Colorado, where Clark now works, plays, and enjoys his mountain home, dubbed, "Rancho Europa."

Our lives are filled with many fascinating things. How did I come to love the myriad of small worlds known as asteroids? First, let me try to help you visualize what these little objects are like, which is not easy. After all, such extraterrestrial environments seem truly alien to beings evolved on an immense, apparently flat landscape. Notwithstanding the failure of Christopher Columbus' ships to fall off the sea's "edge," no amount of book learning compels us to truly feel that our world is round. Perhaps, on a tiny round world, as the French writer Antoine de Saint-Exupery depicted asteroids in his fable *The Little Prince*, we could actually see the horizon falling away just before us and finally know what it's like to live on a sphere. But the laws of physics tell us otherwise. Many years ago, I heard MIT physicist Phil Morrison explain that the nature of matter itself dictates that the Earth must be round, but that rocky asteroids need not be round at all. And—except for a little guy named Dactyl, whom we'll meet much later—they aren't!

Small worlds

As a kid, I toiled for weeks with my crayons to transform my father's desk-top globe into a giant, flat map, drawn on the white sides of dozens of cardboards from his laundered shirts. Immersed in my project, it was just a small leap of faith to picture myself on the

Figure 4.1

The "Little Prince" strides upon his small, round asteroid.

surfaces of the planets, for instance as painted by Chesley Bonestell in *Life* magazine. I also marveled at an occasional comet, hovering over the elms in the smoggy skies of the industrial eastern city that was my home. I was especially amazed at the rings of Saturn, revealed through the cardboard-tubed 2-inch telescope my father assembled. The rings were flat, solid objects, so far as I could see.

How much more difficult it is to visualize a swarm of individual small worlds! That, it turns out, is the reality of Saturn's rings—the Voyager spacecraft's radio experiment showed that much of the bulk of the rings is in house-sized objects. And, despite the representations in sci-fi films and "Star Trek," the asteroid belt is even more difficult to envisage than the rings. The belt is really a torus, located beyond the orbit of Mars, containing hundreds of worlds larger than Rhode Island. It contains perhaps a couple of thousand bodies each having a volume exceeding the volume of water in all of the Great Lakes combined. Yet, if like the "Little Prince" you could venture to the middle of the main asteroid belt, you would barely see a few other asteroids—and they would be among the faintest stars against the black sky of space. Far from ducking hurtling projectiles as in movies, a visitor to the asteroids would be struck by the wondrous emptiness of interplanetary space.

My childhood reading managed, somehow, to miss Saint-Exupery's sweet tale, which might have triggered in me an early fascination with the mysterious environment of life on a tiny world. Instead, notwithstanding a college classmate's project of photographing an asteroid's motions amid the stars, my developing scientific interests increasingly centered on the atmospheres and solid surfaces of the disparate planets. It would be many years before I would turn, by chance, to the asteroids.

I became particularly interested in the craters of the Moon (which I could see through my backyard telescope), and the question of whether they were volcanos or impact craters. (They never looked like pointy-shaped volcanos to me, but then I didn't know at the time the re-markable variety of volcanic profiles.) After graduating from high school in upstate New York, I got a summer job in Arizona where I was assigned to measure lunar craters, for a catalog, using the best telescopic photographs of the Moon. The work was in the fledgling Lunar and Planetary Laboratory of the University of Arizona, which the indomitable Dutch astronomer Gerard Kuiper had just established after relocating from the University of Chicago.

While still a college student in Cambridge, Massachusetts, I published my first professional paper, about lunar craters, in the *Journal of Geophysical Research* (which I was, much later, to edit). When Mariner 4 flew past Mars in 1964, returning a few grainy, light-struck pictures, I marveled at the immense circular ramparts on the surprisingly cratered landscape of our supposedly canal-crossed neighboring world. Naturally, I wanted to extend my studies of lunar craters to Mars. For that college project, I was fortunate to collaborate with two of the three greatest planetary scientists of my lifetime, Jim Pollack and Carl Sagan. (The third was Gene Shoemaker, to whom I'll return. All three are now, sadly, deceased.) Sagan, the famous popularizer of astronomy, was then formulating his vision of Mars as a wind-blown, desiccated world, inhospitable to the life he so fervently wished were there. Thus, along with Pollack (who applied his talents and insights to the broadest range of planetary research topics of anyone of his generation), Sagan was more interested in how the dust was covering up the Martian craters than in the craters themselves.

Gene Shoemaker, on the other hand, developed the basic concept that craters result from the impact of extraterrestrial projectiles, which literally explode when they strike planetary surfaces at hypervelocities of several to tens of kilometers per second. Shoemaker was an extroverted geologist with infectious enthusiasm and a laugh that could be heard a block away. Having demonstrated, in the 1950s, that Meteor Crater in Arizona has the same geological structure as nuclear explosion craters in the Nevada desert, Shoemaker—rock hammer in hand—would lead hundreds of tours down into "his" crater over the ensuing four decades. By taking them into Meteor Crater and across the volcanic landscapes of northern Arizona, Shoemaker trained the Apollo astronauts and a generation of planetary geologists in the techniques of field geology. For Shoemaker, the association between asteroids and craters was second nature. It would take me, and most planetary scientists, much longer to appreciate the profound connection between asteroidal space debris and the evolution of planetary surfaces … and of any life evolved within planetary ecospheres.

I followed the work of Shoemaker since my early work on the lunar crater catalog. He, along with my summer job big-boss Gerard Kuiper, studied the remarkable photographs of small lunar craters taken by Ranger 7 seconds before it, historically, crashed onto the Moon on July 31, 1964. Shoemaker's early, insightful analysis continues to inform my 1990s research concerning the Galileo spacecraft's photographs of the surfaces of the Galilean satellites of Jupiter. In spring of 1968, on the day that Lyndon Johnson told the nation on television that he would *not* run for a second term (which I watched in the Caltech student union), I had my first one-on-one meeting with Gene Shoemaker. He saw me in his role as Chairman of Caltech's Division of Geological Sciences. I had traveled to Pasadena to visit Caltech so I could decide whether to continue my graduate studies at MIT or instead move to Caltech.

I chose to stay at MIT. But it hardly mattered, for my thesis advisor, Tom McCord, had just gotten his degree from Caltech. His connections with Caltech meant that I spent much of my graduate school career in Pasadena and at the nearby Mt. Wilson Observatory. McCord, a short, ruddy-complexioned fellow who had embarked on a scientific career after years as a truck driver, quickly established a small research empire, occupying parts of two buildings at MIT. One day, he called me into his office and offered a thesis topic to me: I would, he suggested, use his new 24-filter spectrophotometer to measure the colors of asteroids. McCord had an optimistic conviction that astronomers could identify different rocky minerals and ices on the surfaces of distant planets; he hoped and expected that simple color measurements would reveal the tell-tale absorption bands. Other graduate students had been assigned Mars, the Moon, and the rings of Saturn. I would do the asteroids, he suggested. Asteroids? Not planets? OK: I'd follow a new path. It wouldn't have seemed strange to Gene Shoemaker.

Shoemaker, disqualified for medical reasons from achieving his boyhood dream of going to the Moon, led the Apollo geology investigations from back on Earth. The astronauts were his surrogates on the

Moon. The lunar program was cut short after Apollo 17. Already disillusioned with NASA politics, Shoemaker began to use *other* approaches (besides journeying to the Moon) to learn about why craters exist. A Princeton-pedigreed geologist with two decades of practical experience, Shoemaker made a radical decision: he decided to become an astronomer. For him, it was obvious that an understanding of the cratering history of the solar system required knowledge about comets and asteroids—the objects that *make* the craters. So he began, along with Eleanor "Glo" Helin, an observational program at Mt. Palomar, north of San Diego, to discover more about those near-Earth asteroids (NEA's) that have wandered in from the asteroid belt to where the Earth orbits the Sun. NEA's are the only asteroids that can crash into our planet. During the several million years that they last in the inner solar system, some of them eventually find themselves intersecting the orbit of the Earth at the unlucky moment that the Earth (or the Moon) is right there. A crater results.

Meanwhile, during 1970 and 1971, I dutifully used Tom McCord's 24-filter spectrophotometer to measure the spectra of sunlight reflected from several dozen main-belt asteroids. The question was whether we could learn what the asteroids were made of and how they were related to various types of meteorites—the stones that fall from the skies. Although young astronomers are normally beset with cloudy skies and equipment failures that protract their graduate-school careers practically into middle age, I was greeted—night after night—with clear skies over my observing posts at Mt. Wilson and at Arizona's Kitt Peak National Observatory.

In the course of writing my PhD thesis, I concentrated on massaging all of my data and trying to understand what they implied about the composition of asteroids. It wasn't as simple as McCord had surmised— absorption bands were few and weak and didn't lead to a straightforward identification of the minerals. Almost as a sidelight, I devoted one chapter of my thesis to cratering of asteroid surfaces. Would they, indeed, be cratered? Despite the tiny gravities on asteroids, could they nevertheless retain regoliths (a term coined by Shoemaker to refer to "soils" on the Moon generated by the repeated crushing and reworking by meteoroid impacts)? Dusty regoliths might affect the colors of asteroid surfaces, modifying the pure mineral spectra. As I thought about asteroidal cratering, it began to slowly dawn on me that there was a direct connection between the asteroids, their own collisional interactions, and the "holes" on planetary surfaces—the lunar and Martian craters that I had been measuring during the 1960s.

Shoemaker's collaborator Helin, a hefty blonde geologist, bore the nickname "Glo," aptly reflecting her generous, outgoing nature. At various scientific conferences during the mid-1970s, Glo tried in her almost messianic way to enthuse me about NEA's. It wasn't that I disagreed that they were interesting and important. I would, much later, devote much of my scientific research energies to NEA's, but— back in the mid-seventies—I still didn't quite get it. I continued to pursue my studies of the mineralogy of chiefly main-belt asteroids and the frustrating question of exactly what mixture of minerals was implied by the various absorption bands in their spectra. I didn't discriminate against NEA's: I observed whatever near-Earth asteroid or "dead" comet nucleus might be available in the skies, and I even obtained some of the best spectra of the large NEA, 433 Eros. Eros, the target of a 1999 visit by the Near Earth Asteroid Rendezvous (NEAR) mission, is about 35 kilometers long. Should it crash into the Earth, which it very well could do sometime in the next few million years, its impact would dwarf even the dinosaur-killing event 65 million years ago. Fortunately, Eros can't harm us during the next few centuries!

Following my 1972 PhD, I continued to measure asteroid spectra, collaborating with McCord and others in his group. My frigid nights in cold domes on Mt. Wilson, Kitt Peak, Anderson Mesa (near Flagstaff), and Hawaii's Mauna Kea gradually evolved into pleasanter nights in observatory "warm-rooms" as observational astronomy evolved into an ever more automated endeavor. Finally, the telescopes and instruments could be left alone in the cold. Meanwhile, NEA searches remained a largely unfunded scientific backwater. The few observers trying to find NEA's operated on shoestring budgets. Geologist-turned-astronomer Gene Shoemaker, later accompanied by his wife, Carolyn, and amateur-turned-pro David Levy, continued to observe into the 1990s in the obsolete, unheated dome of Mt. Palomar's Schmidt telescope—searching for Earth-approaching asteroids, and comets beyond.

An encounter, and a goodbye

It was June 28, 1997, twenty-nine years after I first met Gene Shoemaker. At a patio restaurant overlooking the lake in downtown Columbia, Maryland, I was having a comfortable lunch with Gene and Carolyn. With me were David Morrison and Alan Harris, two prominent asteroid researchers, and Carolyn Porco—leader of the imaging team of NASA's Cassini mission, which was to be launched toward Saturn a few months later. The five of them were the guests to an "encounter"—one of those periods of frenzied scientific and media activities when a spacecraft flies past a body, returning a flood of new data. Encounter activities this time were centered in Laurel, Maryland, at the Applied Physics Laboratory of Johns Hopkins University, the organization responsible for designing, assembling, and operating the NEAR (Near Earth Asteroid Rendezvous) mission. I am one of the select group of scientists involved in overseeing NEAR's camera and spectrometer.

We were basking in the glow of the first incredible pictures returned the previous afternoon from the NEAR spacecraft of the huge black behemoth called Mathilde. The enormous, gaping craters that dominated the form of this dark asteroid (see Figure 4.2) astonished us all. Mathilde looked like a giant piece of blackened Swiss cheese. Yet Mathilde is but a way-station for NEAR before reaching its main goal: NEAR will orbit the NEA Eros in 1999, the culmination of a goal I first heard Gene Shoemaker advocate fifteen years earlier. The six of us reflected on our many interactions during the previous decades that had finally led to this initial success in the first-ever dedicated spacecraft mission to an asteroid. The conversation turned to reminiscing about Gene's long association with Caltech—beginning as an undergraduate student and culminating only a few years earlier with his eventual retirement as a faculty member. Carolyn Porco, a black-haired New Yorker with traces of attitude and accent from the Big Apple, fondly recalled her grad student days in the department with Gene, where her interests settled onto a different swarm of particles (Saturn's rings). I recounted my day with Gene and Lyndon's farewell.

When we returned to the Applied Physics Laboratory, people were abuzz over NEAR's possible discovery, soon to be discounted, of a satellite in the vicinity of Mathilde. It was my job to somehow shield our savvy guests from the "secret" discovery until the science team could confirm that it was for real. Our team leader, a strong-minded Czech-Canadian astronomer named Joe Veverka, whom I first met in college, was determined not to repeat the *faux pas* of another colleague, Lyle Broadfoot, a quarter-century earlier. Broadfoot, nearly incoherent from a sleepless night working on brand-new data from Mariner 10, made headlines worldwide by "discovering" a small moon of the planet Mercury. He named it after his dog. A day later, Broadfoot realized to his great embarrassment that the supposed satellite was, in fact, a well-known distant star far beyond Mercury. Veverka was adamant that history would not repeat itself. He ushered CNN reporters out the door.

I escorted Gene and the other visitors to a remote conference room, and engaged them in a passionate scientific discussion about the recent pictures of the surface of Jupiter's moon, Europa, obtained by the Galileo spacecraft. At issue was whether Europa is a very active world, with a thin ice layer atop an ocean, conceivably teeming with aqueous life. Gene, a half-year before, had advocated the idea that Europa's surface was a billion years old—"young" in comparison with the ancient surface of the Moon, but probably frozen solid and capable of harboring nothing more than ancient frozen fish if it ever had been an abode of life in our solar system. The latest Galileo photos showed almost no impact craters at all on Europa, I told him. Always willing to adapt to new data, Gene was happy to reconsider Europa as an extraordinarily geologically active body.

Then, in mid-afternoon, the Shoemakers departed for the airport. They had a trip to Nova Scotia, where they would meet with David Levy, and then they were off for their annual resuscitation, and crater explorations, in the outback of Australia. They waved goodbye. It was the last time I saw Gene. He would never return from Down Under. In an auto accident, seemingly as unlikely as a globally destructive asteroid impact on Earth—a topic Gene brought into our consciousness during the 1980s—his vehicle encountered head-on, at high speed, another car on one of the rare curves along a desolate track in the Outback.

First Global Image Mosaic of 253 Mathilde

Figure 4.2a

| **Mathilde** | **Gaspra** | **Ida** |

Gaping craters dominate the shape of 253 Mathilde, as imaged by the Near Earth Asteroid Rendezvous spacecraft in June 1997.

Gaining respect

Let me tell you of my studies of asteroids, from the days of my post-graduate studies with Tom McCord's instrument, to the memorable weekend of NEAR's Mathilde encounter. In that quarter-century, aster-

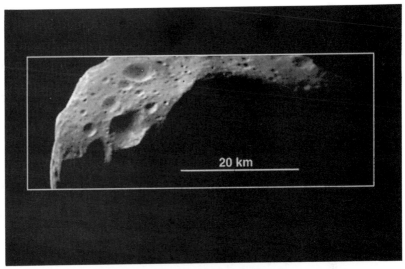

Portion of Closest Approach Mosaic

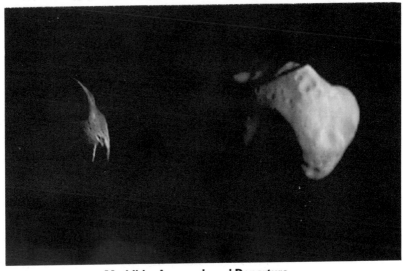

Figure 4.2b **Mathilde: Approach and Departure**

Gaping craters dominate the shape of 253 Mathilde, as imaged by the Near Earth Asteroid Rendezvous spacecraft in June 1997.

oids finally gained respect. They had long been considered the dregs of the solar system. Astronomers had called them the "vermin of the skies" because their trailed images messed up pictures of more glorious celestial objects. Now asteroids are the chief targets of numerous ongoing and prospective spacecraft missions, ranging from a Japanese endeavor to land an asteroid sample in the Utah desert to a would-be first-ever private enterprise deep-space mission, being promoted by Colorado entrepreneur Jim Benson. Asteroids have also become popular icons of the potential catastrophic end of our world—stars of Hollywood blockbusters about the greatest imaginable disasters. Some researchers now think of asteroids as playing one of the most exalted roles in the cosmos: asteroids (and comets) could very well be *the* driving force of biological evolution on Earth, and conceivably everywhere else in the universe, as well.

On top of an office file cabinet, I have a pile of reports about two feet high. With publication dates from the late 1960s to the late 1980s, each one presents the recommendations of some committee or task group about a prospective asteroid spacecraft mission. Except for NEAR, none was ever funded and flown. These committees were assembled most often by NASA, but also by the US National Academy of Sciences, the French and Japanese space agencies, and others. Some reports have plain grey covers, others sport portraits of the Little Prince, and one fancifully depicts a spacecraft orbiting through a hole that pierces a small asteroid. For decades, the accumulating reports gathered dust and all the plans were for nought. But now we are in a new era with missions galore. How did we get here?

My own research—and that of my colleagues—on asteroid reflectance spectra may unfortunately have delayed the exploration of asteroids by spacecraft. It's a story about the hubris of scientists, who are often optimistic about how soon they and their particular techniques can solve crucial problems. As it turns out, there's nothing like going out there to explore a distant world to find out what it's like. Astronomers who study distant stars and galaxies have no choice—our technology will never get us to the stars in our lifetimes. But for asteroids, which look like distant stars through a telescope (hence the word "asteroid," or star-like object), there was another choice. Implementation of the spacecraft alternative was delayed because of competition and expense ... but also because Tom McCord, I, and our colleagues were too successful at promoting our likely success at assaying asteroid mineralogy from afar.

Early this century, asteroids were used by astronomers as standards to calibrate their data as though they were all colorless, gray reflectors of sunlight. But colorimetric studies of some asteroids during the 1950s and 1960s had begun to show that different bodies had different colors. My own spectra went further and revealed a wide variety in the actual spectral features in their reflected sunlight: minerals absorb and transmit light of various wavelengths (colors) very differently from each other, and—evidently—asteroids differ widely in the suites of minerals of which they are composed. See one asteroid and you most assuredly have *not* seen them all!

The more that I, and several other young asteroid researchers, observed these minor planets with a variety of different techniques—mid-infrared radiometry, polarization, and light-curve photometry, which reveals the various shapes and spin rates of asteroids—the more we realized that each of the thousands of known asteroids has its own personality. 324 Bamberga is unexpectedly large, round, and black; and it spins languidly about its axis of rotation, more slowly than the Earth turns. 349 Dembowska is highly irregular in shape, spins around 5 times in a single Earth day, and it has a surface made up largely of the bright green mineral, olivine. 16 Psyche spins even faster than Dembowska, yet has a featureless reflection spectrum like that of pure steel. Some years later, radar echoes bounced off Psyche confirmed that Psyche is the largest known hunk of pure metal in the solar system, 250 kilometers in diameter!

One of the obvious questions for asteroid researchers to address was the connection between asteroids and the rocky fragments, called meteorites, which plummet through the Earth's atmosphere as fiery "shooting stars," lodge in farmers' fields, and wind up in the mineral displays of science museums. Meteorites were once considered to be "thunderstones"—some kind of congealed atmospheric residues formed by lightning bolts. Then, about two centuries ago, several meteorite showers struck in Europe and scientists began to appreciate that these alien, often metal-rich rocks actually come from interplanetary space. Coincidentally, about the same time, the first asteroids were discovered. Putting two-and-two together, a general theory developed that a planet between Mars and Jupiter had exploded, creating the asteroid belt; meteorites were taken to be smaller fragments of the erstwhile planet.

That seemingly plausible hypothesis has not stood the test of time. No large planet ever existed or exploded in the asteroidal region. Instead, asteroids are now known to be the residue of small primordial objects called planetesimals, congealed from the original solar nebula of dust and gas from which the Sun and planets eventually formed. Planetesimals between Mars and Jupiter never managed to form into a planet, most probably because of the gravitational effects of massive Jupiter. Instead of bumping into each other slowly and growing into a planet, the extra velocities imparted by Jupiter resulted in asteroids colliding with each other at many kilometers per second, breaking each other into smaller pieces, resulting in the torus of fragments we see today.

Indeed meteorites are pieces of asteroids, although when I began my observational program back in graduate school, it was very unclear how that could be true. For one thing, nobody could imagine how rocks could be extracted from the asteroid belt, far beyond Mars, so that they could strike the Earth. If you hit a rock hard enough to send it into an Earth-crossing orbit, you would vaporize it. As a start to solving the mystery, I would determine if asteroids were made of the same minerals as the meteorites. Using Tom McCord's spectral reflectance techniques, I would try to determine if certain types of meteorites had the same reflectance spectra as some asteroids. If so, those asteroids *could* be parent bodies for the meteorites. Then we could move on to the question of just what physical processes could deliver meteorites from those particular asteroids.

I must have seemed pretty confident at the time, for the eminent University of Chicago meteoriticist Ed Anders argued that I and my colleagues would soon be successful in linking the meteorites to specific asteroids. Hence, he declared, it would be premature to send a spacecraft mission to an asteroid, for we might simply return a piece of it to Earth only to find that we had merely one more meteorite of an already known kind. NASA was only too happy to send its space-craft elsewhere. During the succeeding decades, however, asteroid researchers have been confounded with puzzles in trying to link meteorites to specific asteroids, even though the extraction and delivery mechanism was eventually solved in the 1980s when it was realized that the dynamics of chaos could gently and efficiently put asteroidal fragments into Earth-crossing orbits.

Like Lewis and Clark exploring northwest America, or Darwin investigating the Galapagos, we early asteroid researchers could do little more than observe, classify, and tabulate. From the mid-1970s to the mid-1980s, such taxonomy was the main activity in asteroid research. By 1974, more than 100 asteroids had been observed with some combination of the spectral, radiometric, and polarimetric techniques. Together, such observations tell not only about an asteroid's color, the shape of its spectrum, and the presence of distinctive absorption bands, but also about how bright or dark the surfaces are. By the mid-1970s, it seemed as though most asteroids were made of moderately reflective rocks rich in pyroxene and olivine, which are the basic minerals (along with quartz, feldspar, and some others) of which the Earth's crust is made. A few asteroids were apparently very different: their surfaces are coal-black in color. Or, so it seemed, until a special collaboration revealed a rather different picture of the asteroids.

In the mid-1970s, I was working at an off-campus research institute in Tucson, near the University of Arizona. Ben Zellner, a stolid southerner, was using Arizona telescopes to painstakingly measure the polarization of light reflected from some of the brighter asteroids. David Morrison, a lanky Harvard-educated astronomer on sabbatical leave in Tucson from the University of Hawaii, observed the infrared heat radiated by asteroids. The three of us compared our data sets and realized that we could classify most asteroids into just two types. The brighter asteroids we termed "S-types," using S as a mnemonic for "silicaceous," meaning composed of the silicate minerals pyroxene and olivine, whose weakened absorption bands were the most prominent features of S-type reflectance spectra. The black asteroids we termed "C-types," using C as a mnemonic for "carbonaceous," since almost all black substances on the Earth (or black meteorites that fall from the skies) are black because of the presence of carbon.

Ten asteroids in our sample were neither fish nor fowl. In our taxonomy, they were unclassifiable, so we called them U's. Naturally, as more asteroids were observed in later years, these early U's became the archetypes for new, less populous classes of asteroids given the letters V, M, E, R, A, and D. Still other distinctive types have been found subsequently, and a veritable alphabet soup of taxonomic types has grown from our meager C, S beginnings.

In our 1975 paper, published in *Icarus* (the international journal of solar system studies, then edited by Carl Sagan), Morrison, Zellner, and I went beyond simple taxonomy to try to understand the population of asteroids. Appearances can be deceiving, and we wanted to understand *all* the asteroids, not just those that were easiest for us to observe. We needed to take account of, and correct for, the biases imposed by our observational circumstances. We realized that the C-type asteroids, being very black and also commonly located in the outer part of the asteroid belt, would be underrepresented in our sample. Compared with a more reflective S-type asteroid in the inner belt, a C-type of the same size would be more dimly illuminated by the faraway Sun, would reflect less light because of its inherently black color, and would be further dimmed by its great distance from Earth. So we would probably miss it, while measuring lots of smaller, closer S-types. We corrected for all of these factors and realized that, despite C-types being only about a third of our sample, the asteroid belt must be overwhelmingly populated by C-types.

Our *Icarus* paper was one of the most widely cited planetary science papers of the decade and helped to make the study of asteroid physical properties a significant subdiscipline of planetary astronomy. By 1975, asteroids had gained enough respect that *Scientific American* commissioned me to write an article about asteroids. A few years later, the National Academy of Sciences started taking seriously the small "dregs" of solar system accretion—the asteroids and comets—and began to draft recommendations about what NASA should do about exploring them. I was invited to join the Academy's planetary advisory committee as the committee began work on its small bodies report. It would turn out to be yet one more report that gathered dust. But work on the report led me into a new direction in asteroid research.

Dinosaurs and impact catastrophes

Looking back, 1980 was a seminal year in melding the study of asteroids with the role of impacts on the Earth and in the solar system. During the 1970s, while I had been observing asteroid physical properties from chilly mountain tops, other observers (especially Gene Shoemaker, Glo Helin, and University of Arizona astronomer Tom Gehrels) were systematically searching for Earth-approaching asteroids. As of 1971, less than a dozen non-cometary objects were known to pass inside the Earth's orbit. By 1979, the number of known Earth-approachers had tripled. By correcting for the great incompleteness of the combined surveys due to inadequate sky coverage and observational biases against faint objects, Shoemaker estimated that the total number of Earth-crossers larger than 1 km across is about 1,300. After two more decades of searching, hundreds have now been found, and refined estimates of the total population have risen slightly, to about 2,000.

Also during the 1970s, NASA's spacecraft were venturing farther afield. Beginning with the surprise of Mariner 4's first pictures of Mars, the "golden age" of planetary exploration was showing that every solid surface in the solar system is being bombarded by comets and asteroids, leaving crater-scarred surfaces. Mariner 10 flew past Mercury in the early 1970s and found that it was heavily cratered, just like the Moon and Mars. In March 1979, the first Voyager spacecraft flew past Jupiter, revealing the cratered surfaces of its two largest moons, Ganymede and Callisto. (It also was obvious why several worlds were virtually crater-free: for instance, Voyager snapped pictures of furious volcanic eruptions on Jupiter's inner moon, Io, that certainly would quickly cover up any impact crater that was formed on its surface. For similar reasons, the Earth's surface retains only the most recent, tiny fraction of impact craters ever formed.)

The intimate connections between craters and small cosmic bodies that had driven Gene Shoemaker through his career gradually became abundantly clear to all planetary scientists, and to NASA. In a summer 1980 meeting in Martha's Vineyard, an elite group of NASA advisors suggested that the space agency look into the question of possible danger from near-Earth asteroids. That same year, the seminal paper was published in *Science* magazine by Nobel laureate Luis Alvarez and his colleagues, who argued that impact of a 10-km-diameter asteroid was the probable cause of the last great mass extinction of species in the Earth's fossil record, 65 million years ago. They had found that a thin layer of rock, enriched in metals (like iridium) that are rare in the crust of the Earth, coincided exactly where fossils of the Cretaceous era suddenly give way to Tertiary fossils (the so-called K/T boundary). Most asteroids and meteorites contain abundant metals; if the iridium-rich layer were distributed worldwide, as was soon verified from measurements of K/T boundary layer rocks acquired from around the world, then a 10-kilometer-diameter asteroid was implicated, according to Alvarez and colleagues.

Although it took another decade of research before gaining general acceptance in the scientific community (and a few doubters remain even today), Alvarez's idea was immediately plausible to planetary scientists familiar with the numbers of comets and asteroids in the inner solar system. One needs to use only simple arithmetic, involving the area of target-Earth and the numbers of Earth-crossing asteroids, to calculate that Earth must be struck by a kilometer-sized object once every 100,000 years, or so. Objects ten kilometers across are almost 1,000 times rarer than kilometer-sized objects, so a body the size estimated by Alvarez should strike the Earth every 50 to 100 million years. The last one evidently hit 65 million years ago: it fits! In fact, it's inevitable that catastrophic impacts have affected Earth's ecosystem in the past ... and will do so in the future.

And that was the theme of a workshop that NASA mounted in summer 1981, in response to the challenge from its advisory council. Gene Shoemaker was asked to chair the meeting, held in Snowmass, Colorado. Along with others coming off the National Academy of Sciences" study of small bodies (our report was published in 1980), I was invited to participate in this "Spacewatch Workshop." I was assigned to a sub-panel of half-a-dozen scientists who examined, from the limited data then available, what environmental and social consequences might result from the impact of asteroids and comets of various sizes. We couldn't second-guess the work of the Alvarez group, who were represented at the workshop, about the effects of an extremely rare 10 km object. But we immediately realized that far more frequent impacts, with far smaller megatonnage explosive yields, might have dire consequences, as well.

Sitting in our break-out room at the Snowmass resort, we realized that impact of an asteroid "only" 2 km across probably would be sufficient to change the climate, destroy agriculture worldwide, and risk the end of civilization as we know it. Our species and most others would survive, but millions or billions of people might die, and we might well be cast into a new Dark Age. We estimated that there was a 1-in-10,000 chance of such a catastrophe happening in our lifetimes. Though small, such chances are greater than dangers that governmental agencies (like the Environmental Protection Agency) and individuals (like frequent air travelers) take very seriously. (Your chances of dying in the immediate aftermath of a cosmic impact are *much* greater than of winning big in a state lottery!) Worse, our panel feared, a much smaller impacting asteroid might be misinterpreted as a nuclear attack, triggering an escalating nuclear war. Remember, we were in the midst of the Cold War then!

Sobered by the discussions, I left the Snowmass workshop convinced that the impact hazard needed further study, and that the military leaders and opinion makers around the globe needed to know more about rare cosmic impacts. I toiled on drafts of the Shoemaker report, but it never saw the light of day. Some suspected that it was squelched because of Cold War sensitivities. So far as I can tell, however, Gene Shoemaker—who always had ten times as many things to do as time to do them—got busy with other responsibilities and NASA officials failed to keep the pressure on him to finish it up. Nevertheless, I felt it was a story that needed to be told and, when David Morrison and I decided to work on a popular trade book on astronomical catastrophes, I had my chance. The concluding chapter of our book *Cosmic Catastrophes*, published in 1989, summarizes the Snowmass workshop and its conclusions about the cosmic impact hazard.

Although our book sold poorly, it was read by some people who count—for instance, by Congressional staffers associated with the space caucus. During the subsequent decade, public awareness of impacting asteroids has gone from nothing to truly bizarre levels: bad TV mini-series now deal with astronomers and Emergency Management officials responding to crashing rocks; commercials show fiery objects descending from the skies; Federal Aviation officials compare airline hazards with the risks of falling comets; and blockbuster movies about deadly asteroids are produced by Hollywood studios. The spectacular crash of Comet Shoemaker-Levy 9 into the planet Jupiter in 1994 had much to do with reinforcing the perceived reality of the danger from the skies ... but that is another story. Meanwhile, although a few modest telescopic searches for Earth-approaching asteroids continue, essentially no NASA funds have been made available to research the impact hazard.

NASA funding notwithstanding, an intellectual revolution is emerging about the role of asteroids in the cosmos. As astronomers search for planetary systems around other stars, they are finding "disks" around some stars. Although most stars are too far away to tell if they have planets let alone smaller bodies, it is likely that some of these disks are composed of swarms of asteroids, comets, and other space debris grinding itself into interplanetary dust. If Earth-like planets exist around such stars, any emerging life would have to deal with the continual rain of asteroids and comets onto their surfaces. It is now believed that evolution of life on our own planet was delayed until the last huge ocean-evaporating, sterilizing impacts had ceased 4 billion years ago. There remain occasional K/T-like impacts to this epoch, which drastically rearrange ecological niches, producing abrupt changes in evolution of species in between long periods of stasis, akin to the punctuated equilibrium model of evolution touted in popular essays by the Harvard evolutionist (and baseball fan) Stephen Jay Gould. University of Chicago paleontologist David Raup has suggested that *all* of the great mass extinctions might eventually be explained as the result of cosmic impact.

Given the existence of huge comets and asteroids still remaining in the solar system today, it is just a matter of chance that one of them—like the asteroid Eros or Comet Hale-Bopp—failed to strike during the 65 million years it has taken mammals (and ourselves!) to evolve from the ashes of the dinosaurs. We may very soon have the technology to colonize other planets and render ourselves immune from an eco-catastrophe on the planet of our birth. Already, we have the telescopes to discover, and the rockets and bombs necessary to divert, any incoming asteroid. The probability of ubiquitous asteroid bombardment throughout other planetary systems in the universe just might explain why we have failed, so far, to detect or make contact with other intelligent civilizations, despite radio telescopic searches. Maybe Earth is unusually well protected from bombardment (Jupiter has been hypothesized to shield us from what would be even greater cometary bombardment) or maybe our evolution proceeded *just* fast enough to give us the chance to protect ourselves before something like Eros hits.

Spacecraft encounters

Through a series of quirks, I've managed to be sitting in the "front seat" during the first three encounters of asteroids by spacecraft. I'm among a lucky few people who got to see, as soon as the data hit the ground, the first close-up images of the asteroids Gaspra, Ida, and Mathilde. And I'm looking forward to the fourth: Earth-approacher Eros. How did this happen?

Well, to be a planetary scientist, you have to write proposals to NASA. There are several slots on the faculties of a few well-endowed or state-supported universities, but planetary science is not a core curriculum subject requiring numerous teachers, compared with traditional sciences like physics, chemistry, and biology. So most planetary researchers live on salaries funded by NASA grants (the National Science Foundation, and other potential funding sources, mainly expect NASA to support planetary science). One can propose for funds to study some particular topic. Or one can propose to participate in a planetary mission. The available funds, and mission opportunities, are few so that competition is stiff and the chances of success for any given proposal are low. (Even as I write this, I've received yet another rejection notice, despite my colleagues' views that my proposal addressed one of the most important questions in planetary science and that I was the right person to do the research.) So one raises one's chances, in order to put food on the family table, by writing lots of proposals. The alternative, of course, is to turn one's talents toward more lucrative endeavors. But I, like most planetary researchers, love the field. And we owe it to the taxpayers, who have invested so heavily in our schooling and our experience, to continue trying to apply our hard-won expertise to the studies we were trained to do—whatever short-term funding shortfalls exist.

So it was, in the mid-1970s, that I wrote a proposal to participate in the Galileo mission to Jupiter. Although Jupiter was not my first love, I had accumulated a credible list of publications about Jupiter and I could speculate about craters on Jupiter's moons. Apparently, I was one of three "youngsters" added to the Galileo imaging team as a risky experiment, along with a group of more experienced gray-beards. As someone now sporting a gray beard myself, I can testify that Galileo has been the project of a lifetime. It is still going on! There were delays in developing the Shuttle—cancellation (due to funding cut-backs) of an energetic upper stage; the Challenger explosion; an extraordinarily long billiard-ball-like path through the solar system—until, eventually, Galileo reached Jupiter in late 1995, thirteen years late. That, too, is a story for another occasion.

However, Galileo's long traverse through the solar system, combined with recognition by NASA officials that asteroids were getting the short end of the stick (no approved missions through the early 1990s), resulted in a piece of enormous luck. So that NASA could "check off" asteroids in its planetary exploration matrix, the Galileo Project was mandated to examine a couple of asteroids. It was an unfunded mandate: for no additional cost, Galileo managers were instructed to fire its thrusters in order to send it very close to two asteroids already along its path to Jupiter. Moreover, they would turn on the spacecraft's instruments and return data to Earth. Thus, by luck, the one spacecraft mission I, as an asteroid specialist, was then involved with—a Jupiter mission—would become the first to take pictures and spectra of aster-oids!

I vividly recall the moment on the morning of November 9, 1991, sitting in a room in Jet Propulsion Laboratory's Multimission Image Processing Laboratory, when the first image of an oddly angular asteroid, named 951 Gaspra (see Figure 4.3, color section), showed up on the television screen. My colleague, Cornell geologist Peter Thomas, immediately began sculpting away pieces of a gridded sphere on his computer console to create a realistic three-dimensional representation of Gaspra's shape. Throughout the day, the pictures continued to be radioed back, ever so slowly, from the distant spacecraft, which had zipped past Gaspra two weeks before. The engineers had worked out an ingenious method of dribbling back a sampling of data from Galileo (whose main antenna was dysfunctional), rather than waiting a year for the spacecraft to return to Earth and disgorge its tape-load of data.

Gaspra certainly didn't look like any of the spherical worlds of the Little Prince. Nor did it exactly resemble a chip of a rock. Its shape, the most irregular of anything yet photographed in the solar system, was uniquely its own. To some of us, it looked almost as though two objects had joined into one. To others, Gaspra appeared to have had sections of its outer surface sheared off by glancing impacts. Immediately, something seemed odd to me about Gaspra—its surface is peppered with numerous, small craters, but it lacks big craters, and the few moderate-sized craters it does have hardly cover its surface: it is not at all like the Moon, which is saturated by overlapping craters. To the degree that the numbers of craters of different sizes reflect the small-impactor population, we had our first glimpse into the numbers of small main-belt asteroids some tens of meters across.

Once all of the Gaspra data were back, following Galileo's Earth encounter in November 1992, I earnestly began to measure the craters on Gaspra, as I had measured lunar craters 30 years earlier at the beginning of my career. I wanted to do a comprehensive job. By the summer of 1994, hoping to finish my manuscript, I carried my Gaspra papers in my laptop computer bag to a scientific conference in Prague. But my bag was swiped by a pair of thieves in the train station. It took more than a year before I could re-do and finish the work; my paper finally saw the light of day in *Icarus*, along with all the results about Ida, in 1996.

Gaspra was but a foretaste compared with the wonders of Ida (see Figure 4.4, color section), encountered by Galileo on August 28, 1993. At closest approach, the much larger Ida filled up five separate frames of Galileo's camera. Galileo took pictures of Ida through color filters equivalent to those I had used back in 1975 when I observed Ida from Hawaii's Mauna Kea Observatory. As a result, pieces of a puzzle that had eluded us for a quarter century began to fall into place. Our color composite pictures showed that small, recent craters on Ida have different colors from most of Ida's surface. Apparently, the recent craters had dug into fresh bedrock revealing Ida's "true" colors. But after a spell of time, some process (perhaps the impacts of micro-meteorites) was changing the spectral character of Ida's surface. No wonder it was difficult to directly match laboratory spectra of meteorites to the asteroid spectra I had struggled to obtain! Exposure of Ida's regolith to space seems to depress the apparent strength of the mineral bands and gives Ida an overall reddish tint—just the traits that had seemed to invalidate a connection between the S-type asteroids (like Ida) and the most common meteorites, the so-called ordinary chondrites. Ordinary chondrites are believed to be heated-but-never-melted samples of the original material accreted from the solar nebula 4.5 billion years ago. Mysteriously, these common and important meteorites had not been previously linked to any spectral type represented in the main asteroid belt. Now that we see how "space weathering" processes can change mineral spectra of ordinary chon-drites to look like S-type asteroids, I have renewed hope that we can connect other meteorite types with asteroid types.

Ida continues to fascinate me. More convincingly than Gaspra, Ida's shape makes it look like the merger of two separate asteroids. Unlike Gaspra, it is covered with craters. Either it is much older than Gaspra, or else Gaspra is much stronger than Ida (for example, made of metal) so that it is difficult for impacts to gouge out craters in Gaspra's surface. Also unlike Gaspra, Ida sports a small mile-wide companion—a little spherical object orbiting around Ida, named Dactyl. It is unusually spherical, a fine place for the Little Prince, and a puzzle: after all, as Phil Morrison had explained, the natural shape for something so small is irregular. Dactyl's existence also raises the question of whether other asteroids have satellites, and—if so—how they come to be formed. For now, the Ida/Dactyl system is yet another case of each asteroid having a different personality.

By the early 1990s, I had temporarily given up serving on NASA committees and I was getting disillusioned with the proposal-writing rat-race—ever more proposals chasing ever smaller slices of a slowly dwindling pie. So I let some proposal opportunities pass by, including the Cassini mission to Saturn. But one day, at a June 1994 conference in Flagstaff, Arizona, University of Texas astronomer Anita Cochran mentioned to me that she was proposing for the Near Earth Asteroid Rendezvous mission, and she wondered about how my own proposal was going, due in a few days. Truthfully, I hadn't even thought about proposing, even though—years earlier—I had participated in the work of the NASA committee that had drafted the mission concept. (This, in fact, was the report with the spacecraft flying through a hole in the asteroid on its cover.)

I returned to my office and spent a single day dashing off a proposal for NEAR. I explained how I would help design the infrared experiment and interpret the spectral data, using my experience with spectral observations of asteroids. I didn't propose for NEAR's imaging team, which I guessed would be oversubscribed. A few months later, I was astonished to receive a phone call at home, telling me that my proposal was accepted. Moreover, I was told, the infrared-spectroscopy and imaging teams would be combined—I'd have responsibilities for both investigations! Thus I became part of the Science Team of mankind's first *dedicated* mission to an asteroid. The proposal-writing process reminds me of my college days when it didn't matter whether I struggled over an assignment for weeks, or dashed off a paper during the half-hour before class began. My humanities teaching assistant instructor, Erich Segal (later to become the famous writer of *Love Story*), consistently graded my papers no higher than B- and no lower than C+. Despite all the toil and agony, NASA Proposal Review Panel funding decisions often end up looking like flips of a coin.

NEAR is one of those fast-track missions in NASA Administrator Dan Goldin's cheaper/faster/better mould. The spacecraft and instruments were largely built before the new science team ever met. A little over a year later, we watched a Delta rocket propel the Volkswagen-sized spacecraft into the skies above Cape Canaveral. Like Galileo's investigations of Ida and Gaspra, the NEAR project availed itself of a chance opportunity: Mathilde was along NEAR's path toward Eros. So it was, a little over a year after launch, that I found myself (along with Gene and Carolyn Shoemaker, and our other guests) staring at images of another asteroid with its own unique personality: Mathilde. Even Gene Shoemaker, who had studied more craters than anyone else on Earth, had never seen craters so (comparatively) large! Of course, there *are* larger craters on much larger bodies and planets. But for modest-sized Mathilde, only 50 kilometers across, to sport not just one but *several* craters exceeding 25 kilometers in diameter was unbelievable. Mathilde's whole shape is an amalgamation of the ramparts of giant craters.

The NEAR Science Team struggled to understand how Mathilde's unique appearance might be a clue to another special trait, recently appreciated from analysis of telescopic photometry of this black, C-type asteroid: it hardly spins at all. Tumbling ever so slowly in space, Mathilde gradually turns around every two or three weeks, each time about a slightly different axis (there are no fixed north or south poles on Mathilde). Nobody could imagine why this might be so, so we hoped our pictures might tell us. But apparently not. And all of NEAR's other instruments, which might have provided different clues, were not yet turned on. After all, NEAR is a dedicated mission to the Earth-approacher Eros, not to faraway Mathilde, where the sunlight is so dim that NEAR's solar panels could power only a single instrument, the camera. Maybe detailed analysis of the flyby pictures will eventually yield clues to Mathilde's slow, tumbling spin; otherwise, Mathilde will hold her secrets, for it will be a long time before we're likely to get back to this serendipitously explored small world.

As NEAR sails on toward its rendezvous with Eros, I can only marvel at the variety of small worlds we have already seen among the asteroids. For me, asteroids benefit especially from exploration by spacecraft. From Earth, asteroids are mere "points-of-light." Astronomers struggle to milk every bit of information gleaned from photometry of these small remote worlds and still barely scratch the surface. As Ida, in particular, taught us, the *average* light from an asteroid hides essential details that require spatial resolution—like the distinctive colors of small, recent craters. And no amount of theorizing or computer modelling can replace up-close exploration, just going out there to *look*, which is why we are so often surprised by pictures returned from deep space.

NEAR's mission director, Bob Farquhar, is a lovable, puckish, eccentric soul, whose talents at trajectory wizardry have only gradually gained overdue respect. His latest dream is to end NEAR's explorations in the year 2000 by dropping it directly and gently onto the surface of Eros. If this daring end-game succeeds, we may see pictures of small, individual rocks on Eros—which would be an astonishing improvement over the football-field scale of resolution of asteroid images obtained to date. What a remarkable achievement it will be to gather comprehensive, detailed maps of what appears from Earth to be a just a slowly moving star, too faint even to be detected by a small telescope let alone by the naked eye. As we study Eros, let's remind ourselves of its very real potential, several million years from now, to transform the biosphere of our planet more dramatically than the K/T impactor did on the day the dinosaurs died. Perhaps, by that epoch, our species will be exploring asteroids around distant stars and may even have encountered other intelligent species lucky enough to have evolved and developed technology in time to escape the random cosmic bombardments that help shape the origin, evolution, and destiny of life throughout the universe.

Outer Worlds

5

Cruisin' with Comet Halley

PAUL WEISSMAN

Paul Weissman. Mr. Comet, Mr. Oort Cloud. That's Paul Weissman. A Brooklynite who migrated west to California in the early 1970s, Paul did his graduate work at UCLA (University of California at Los Angeles), and then stayed in the LA area to do research at Caltech's Jet Propulsion Laboratory (JPL), the home of most of NASA's planetary exploration projects. Since he's been there, Paul (a theorist) has produced scientific papers and reviews on all aspects of cometary physics and dynamics, as well as over 30 popular articles in planetary astronomy, and one children's book. He has also been involved in virtually all of the major NASA comet missions that have been planned or flown. In the last two years, Paul has begun some ground-based comet observing, and in an incredible stroke of luck found himself one recent evening giving a personal tour of McDonald Observatory's venerable 82″ telescope to Mick Jagger and family! Paul is one of the most humorous planetary scientists (and one of the best speakers) in our field. He's also a mean poker player, a Miata owner, and about the best bachelor uncle to little kids you're likely to find. After 25 years of comet studies, Paul tells us in the essay below how he got into this line of work, and why comets are, in his view, about the best research subject one could ever hope to find.

My first encounter with Halley's Comet came during my PhD thesis exam in 1977. For about two hours I had been carefully describing the intricacies of the physical and dynamical evolution of long-period comets to my committee of five UCLA professors. Suddenly, George Abell, a renowned galactic astronomer, wanted to know about Halley's Comet. George approached everything astronomical with gusto and enthusiasm. He was the author of the most widely used undergraduate textbook in astronomy, and was well known for another enjoyable sideline, lecturing on luxury cruise ships that steered themselves into the path of total solar eclipses.

"When will the comet come back?" he asked. "How bright will it be?" "Will there be spacecraft missions to the comet?" In the back of my mind a little voice pointed out that my thesis had almost nothing to do with Halley's Comet. So George's questions probably meant that the exam was over and that I had passed. As the little voice started screaming with joy, I struggled to remain outwardly calm and to field George's flood of questions.

"Where's the best place to go to see the comet?" George asked. That was enough for the other faculty members who had places to go and things to do. "You can't send a cruise there, George!" grumbled Bill Kaula, the committee chairman, in his dry New England style. Everyone laughed. With that, the exam was over and I was a real astronomer.

Sadly, George Abell did not live to see the return of Halley's Comet nine years later. If he had, he would have surely enjoyed the immense public interest in the comet, and the opportunity to educate the public on the nature of these fascinating and beautiful visitors from afar. And he would have seen Bill Kaula's comment proven wrong.

Halley's Comet is a legendary body. Its returns have been observed like clockwork every 76 years for over two millennia. It was seen by Chinese astronomers in 240 B.C. and 17 orbits later hung over England just before the Battle of Hastings in A.D. 1066. The renowned American writer Mark Twain commented that he had come in with Halley's Comet in 1835, and he successfully predicted that he would go out with it in 1910. More importantly for astronomers, Halley's Comet was the first comet recognized as periodic, that is, the first comet which was recognized to be in a regular orbit around the Sun that would bring it back at predictable intervals.

Edmond Halley was one of those renaissance scientists with an insatiable curiosity and an incredible breadth in the things that interested him. As a student he spent a year on St. Helena in the South Atlantic charting the stars of the southern sky. Later he would sail twice to the New World to investigate the Earth's magnetic field, and to chart the circulation of its atmosphere and oceans. As secretary of the Royal Society, he gave up his salary one year to pay for the publication of the work of a reclusive mathematician and physicist named Isaac Newton—the book, Newton's *Principia*, is one of the most important books in science. Halley also worked out the first mortality tables for life insurance companies and helped develop one of the earliest diving bells. Halley himself is described as a very affable and outgoing fellow, with a fondness for brandy and an ability to "swear like a sea dog," both acquired during his sea voyages to the New World. But what he is best known for is his work on comets.

Halley and his friend Newton were interested in applying the new Law of Gravitation to understanding the motion of heavenly bodies. Aided by the Laws of Planetary Motion, earlier coined by Johannes Kepler, Halley and Newton used observations of several different comets and solved for each comet's orbital elements. Orbital elements are a mathematical description of the path of a planet or comet or asteroid around the Sun, or a satellite around its planet. They describe the average distance of the object from the Sun (called the semimajor axis), how much the orbit resembles or departs from a perfect circle (called the eccentricity), how much the orbit is inclined to the plane of the Earth's orbit (the inclination), and so on.

The reasons for Halley's interest in comets are obvious. These beautiful objects with their long glowing tails would appear at random and traverse very different paths across the sky from those of the known planets. Early speculations that comets were clouds of glowing gas high in the Earth's atmosphere were easily disproved. (If a comet was within the atmosphere, then two observers at different points on the Earth's surface would see the comet projected against different stars in the sky—when this experiment was actually done by Tycho Brahe in 1577, it was found that this was not the case.) But what then were the comets and where did they come from? Halley hoped that, by determining their orbits, he would help to answer those questions.

Halley had observed a number of bright comets himself, including one in 1680 that passed extremely close to the Sun, and another bright one in 1682. But most of the observations he used came from other astronomers, dating back over several centuries, even before the invention of the telescope. He eventually collected data for 24 comets, and set about finding their orbital elements. A copy of the comet catalog from Halley's classic paper, published in London in 1705, is shown in Figure 5.1.

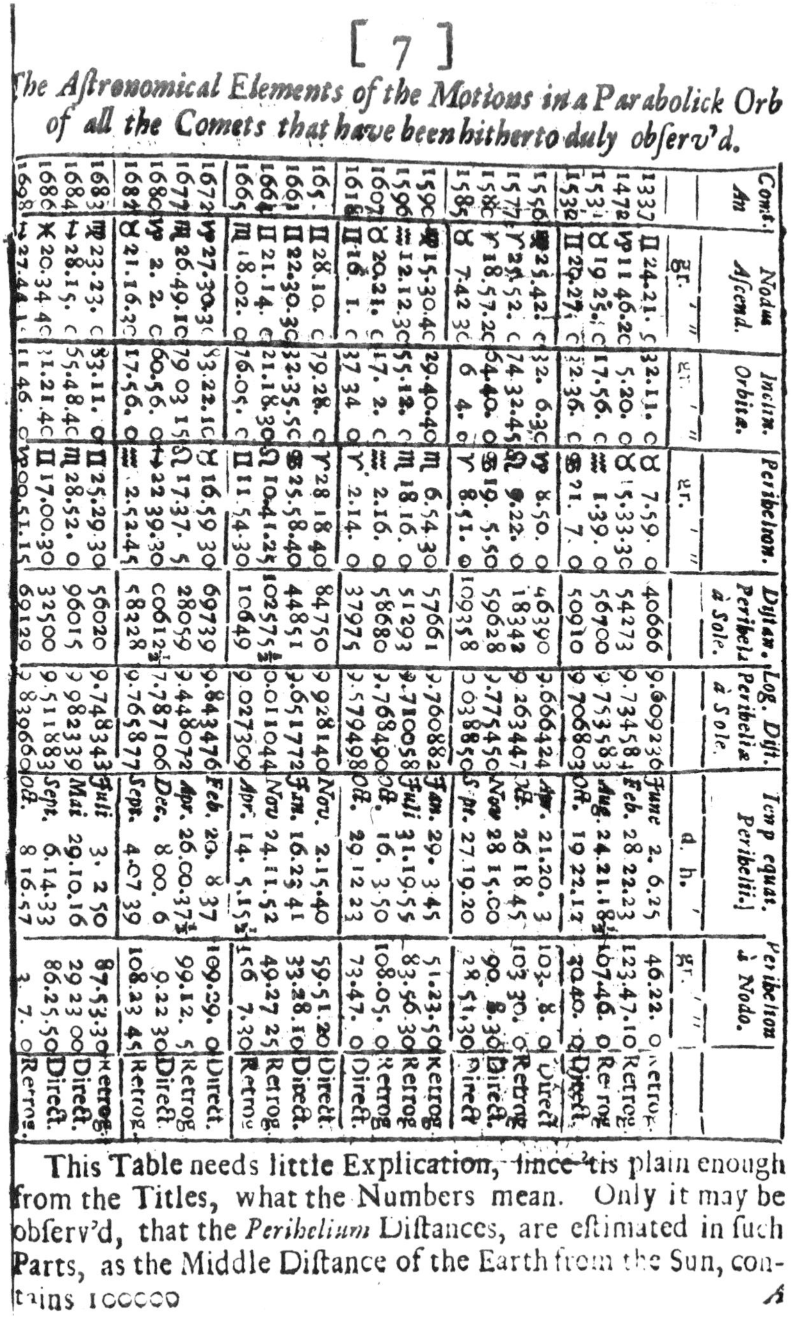

[7]

The Astronomical Elements of the Motions in a Parabolick Orb of all the Comets that have been hitherto duly observ'd.

Cometæ An	Nodus Ascend.	Inclin. Orbitæ	Perihelion.	Distan. Perihelia à Sole.	Log. Dist. Perihelia à Sole.	Temp. æquat. Perihelii.	Perihelion à Nodo.

(A table of 24 comet orbits with numeric astronomical elements follows; the columns list the ascending node, inclination of orbit, perihelion, perihelion distance from the Sun, its logarithm, the equated time of perihelion, and the perihelion from the node, for comets from 1337 through 1698.)

This Table needs little Explication, since 'tis plain enough from the Titles, what the Numbers mean. Only it may be observ'd, that the *Perihelium* Distances, are estimated in such Parts, as the Middle Distance of the Earth from the Sun, contains 100000

Figure 5.1

The catalog page from Halley's historic paper "A Synopsis of the Astronomy of Comets," published in London in 1705. There are 24 comet orbits in the table. Comet Halley appears three times, in 1531, 1607, and 1682.

Unlike the planets whose orbits are close to circular, Halley found that the comets moved on near-parabolic orbits. Parabolic orbits are open, that is, the object does not follow a closed path around the Sun. Instead, the comets appeared to pass only once through the planetary system and then to head back out to interstellar space. Unfortunately, the poor quality of the observations prevented Halley from determining the orbits any more precisely.

But Halley also noticed something very interesting in his catalog. Three comets, those of 1531, 1607, and 1682 (the one Halley had observed himself) had remarkably similar orbits. Their closest distance to the Sun was each about six-tenths of the average distance of the Earth from the Sun, and their orbits were each inclined 162 degrees to the Earth's orbit. This high inclination meant that these comets actually traveled around the Sun in the opposite direction from the Earth's motion. This is known as a retrograde orbit. And the three comets were spaced at roughly equal intervals of 76 and 75 years. To Halley, these comets seemed too similar to be just a coincidence.

Halley reasoned that these were not three different comets, but rather the same comet returning regularly every 75–76 years. He predicted that the comet would return again in 1758. Since he was 49 years old at the time, he also predicted that he would not live to see the comet return (remember those actuarial tables which he devised!). Halley asked that, if the comet did return, he hoped that it would be remembered that the prediction was made by an Englishman.

Both of Halley's predictions proved correct. He died in 1742 at the age of 85, 16 years before the comet was found again by a German amateur astronomer, Johann Palitzsch, on Christmas day, 1758. Halley was right! Comets did indeed move on regular orbits around the Sun. To honor Halley and his brilliant prediction, the comet was given his name. Halley's Comet returned again in 1835 and again in 1910.

The 1910 appearance was particularly exciting because the comet came very close to the Earth, only a mere 9 million miles away (trust me, this is close by astronomical standards!). The Earth actually passed through the tail of the comet. This caused some concern at the time because only a few years earlier it was discovered that comets contained cyanogen, a poisonous gas. Would the comet kill everyone on the

Earth? As the world held its collective breath, absolutely nothing happened.

The Oort cloud

Halley's Comet would return again in 1986. But it was still out beyond the orbit of Uranus in 1973 when I was looking for a topic for my PhD thesis at UCLA. My research advisor, George Wetherill, a world-renowned geochemist, suggested that I consider studying the Oort cloud. "What's that?" I asked, never having heard the term before. "It's a hypothetical cloud of comets surrounding the solar system," replied George. "It's an idea proposed by Jan Oort in 1950 but no one has been able to show that it really exists."

Oort was a name I recognized. Jan Hendrik Oort was a brilliant Dutch astronomer who had determined the rotation of the Milky Way galaxy in the 1920s. I had heard him speak at an American Astronomical Society meeting in 1971, where he was introduced as the greatest astronomer of the twentieth century. But what did he have to do with comets?

In the late 1940s Oort had become interested in the origin of the long-period comets through several colleagues who were determining comet orbits and studying their dynamics at Leiden Observatory in the Netherlands. As Halley had found, most comets approached the planetary system on very eccentric, near-parabolic orbits. Such orbits can have very long periods, a million years or more. Comet Halley and other comets with periods less than 200 years are classified as short-period comets; the definition is somewhat arbitrary and is based on the fact that good observations are only available for the past 200 to 300 years. Comets with orbital periods greater than 200 years are called long-period comets.

As observations and data reduction methods improved in the nineteenth and twentieth centuries, it was possible to determine far more precise orbits for the long-period comets, much better than the simple parabolic solutions found by Halley and Newton in the early 1700s. Looking at the improved orbits, astronomers found that about one-third of the long-period comets appeared to be on hyperbolic orbits.

Hyperbolic orbits are not bound to the solar system, and come from interstellar space. The other two-thirds of the orbits were mostly very long-period ellipses, gravitationally bound to the solar system, but with many extending out halfway to the nearest stars, and with orbital periods of many millions of years.

To understand better where the long-period comets were coming from, astronomers began to integrate the orbits backwards in time. Just as the future motion of a comet can be predicted by carefully calculating its orbit and the tugs of gravity from each of the planets, the past motion of the comet can be reconstructed by simply inserting negative times into the equations of motion. When this was done, the astronomers found that the comet orbits were so weakly bound to the solar system that the distant gravitational tugs, or "perturbations," of the planets could change the comet orbits significantly as they moved in towards the Sun. When the orbits were traced backwards to a time before the comets entered the planetary system, and then referenced to the center of mass of the entire solar system and not just to the Sun, it turned out that virtually all of the apparently hyperbolic orbits were, in fact, elliptical. Thus, it was shown that the long-period comets were all members of our solar system.

But the distribution of the cometary orbits was not uniform. When plotted as a function of the orbital energy (see Figure 5.2), it was found that about two-thirds of the comets were distributed in orbital energy in a low, flat distribution. But at the very weakest orbital energy, corresponding to orbits which traveled 20,000 AU[1] from the Sun or more, there was a large spike of comets. Such orbits are at the very fringes of the Sun's gravitational field of influence, where the gravitational tugs from other stars are almost as strong as that from the Sun. Why were so many comets apparently coming from so far away?

A colleague of Oort's at Leiden Observatory, A. E. E. van Woerkom, showed that the low, flat distribution was the result of distant perturbations by the giant planets, primarily Jupiter. As a comet passed through the planetary system, the random gravitational tugs of Jupiter

---

[1] An AU, or astonomical unit, is the average distance of the Earth from the Sun.

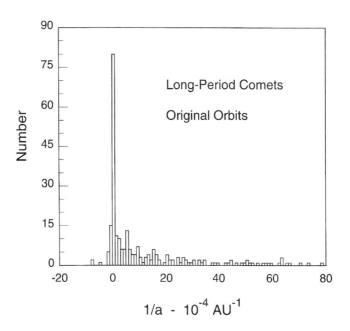

Figure 5.2

$$1/a - 10^{-4} \text{ AU}^{-1}$$

Histogram of the number of long-period comets versus orbital energy. The orbital energy of a comet is given by GM/a, where G is the gravitational constant, M is the mass of the Sun, and a is the mean distance of the comet from the Sun, called the semimajor axis. Since G and M are constants and are the same for all comets, the orbital energy is simply proportional to 1/a, the inverse of the semimajor axis. In the histogram, the low continuous distribution at the bottom of the plot represents the returning long-period comets which have passed through the planetary region before. The comets in the spike are "dynamically new' comets, making their first pass through the planetary region from the Oort cloud. The few comets at negative values of 1/a appear to be in hyperbolic orbits. However, these are probably the result of small errors in the observations or in the determination of the orbits.

and the other planets would change the energy of the orbit, modifying it to either a smaller, more tightly bound ellipse, or a larger ellipse, more loosely bound. The planetary perturbations could even eject comets to interstellar space on hyperbolic orbits. But what then was the spike of comets at near-zero energy?

Oort recognized that the spike had to be the source of the long-period comets. Somehow comets were being fed into the planetary system from very far away, but from a source that was still part of the solar system. Oort suggested that the planetary system was surrounded by a vast cloud of comets stretching halfway to the nearest stars.

The comets in the cloud were so weakly bound to the Sun that their orbits were perturbed by random passing stars. As the Sun and solar system moved in its orbit around the galactic nucleus, stars would randomly approach close enough to gently pull on the comets and change their orbits ("close" in this case means stars passing within about 200,000 AU of the Sun—about 5,000 times the distance of the most distant planet, Pluto, from the Sun!). These gentle tugs are enough to throw the comets into planet-crossing orbits where we can observe them.

In order to explain the number of observed comets, Oort estimated that the cloud had to contain 1.9×10^{11} comets. That's 190 billion comets! The orbits of the comets extend out literally halfway to the nearest stars, 100,000 to 200,000 AU from the Sun. Compared with these distances, the planetary system is actually a very small and very far away target to shoot comets at (the orbit of the most distant planet, Pluto, is only ~ 39.5 AU from the Sun, on average) and that is why so many comets were required.

Where did the comets in the cloud come from? Oort suggested that the comets had been ejected out of the asteroid belt. But Oort did not know that comets were icy bodies, and that the asteroids contain little or no ice. That same year, 1950, Harvard astronomer Fred Whipple pointed out that many of the observed features of comets could be explained if the comets were icy conglomerates, big "dirty snowballs" in space, a few miles across. Whipple suggested that as a comet approached the Sun its ices would sublimate (go directly from the solid to the vapor state), and the outflowing gases would carry with them grains of fine dust frozen in the cometary ices. These gases and dust would produce the glowing head or "coma" of the comet, and then be blown back by the pressure of sunlight and by interactions with the solar wind to form the beautiful cometary tails.

Also that same year, 1950 (it was a very good year for comets!), astronomer Gerard Kuiper at the University of Chicago pointed out that the icy nature of comets required that they had formed farther out in the planetary system, among the giant planets where it was cold enough for ices to condense. In fact, the comets (along with the asteroids) were the first macroscopic bodies to form in the primordial solar nebula, the huge cloud of dust and gas out of which the Sun and solar system formed. Ice and dust grains were slowly brought together in the solar nebula as material fell inward toward the young, forming Sun. As these clusters of ice and dust ran into each other, they stuck, growing larger and larger bodies. When all the ice and dust was swept up in this manner, the comets had formed. But then the comets started running into each other and growing even larger bodies. Finally, several of these bodies grew so large that they began to gravitationally capture the gas in the solar nebula, and these objects became the giant planets Jupiter, Saturn, Uranus, and Neptune.

But even after the giant planets formed, there were still lots of comets left over, wandering the outer planetary system between the orbits of the four giant planets. The planets had grown large enough that they could gravitationally fling these comets into very distant orbits, or even to interstellar space. As more and more comets were pumped up to very large orbits, the Oort cloud was slowly populated.

All of this was still only a hypothesis when I began my thesis work in 1973. Both Oort and Whipple had constructed simple dynamical models to try to verify the hypothesis of a distant cometary cloud. A dynamical model is a mathematical simulation, often run on a computer, which allows a scientist to examine what might happen under a combination of physical laws, such as the all-important Law of Gravity, and various assumed starting conditions. Oort and Whipple's simple models had shown that the idea of a cometary cloud made good sense, but there were still many doubters.

And so I set about the task of building a comprehensive, computer-based model of the evolution of long-period comets from the Oort cloud. Such models are often called Monte Carlo models, after the casino at Monte Carlo. They assume that many physical and dynamical processes are dominated by random chance, such as the random chance that a star will pass close to the solar system, or the random chance that a comet will pass close to a planet. By applying the laws of statistics, one can build a model in the computer that simulates the real world.

In my model I tried to include everything that we knew could affect the long-period comets. I included gravitational perturbations by the planets and by random passing stars. I included the possibility (thankfully small) that a comet could run into a planet, or into the Sun (both of these have actually been observed!). I included the fact that comets occasionally break apart for no apparent reason as they approach or move away from the Sun, and the possibility that a comet might sublimate away all of its volatile ices and disintegrate. Finally, I included the fact, pointed out by Whipple, that the jetting of gas and dust off the surfaces of comets acts like a little rocket engine and actually manages to change the orbits of the comets slightly.

By running my model in the computer with different starting conditions and different probabilities for random disruption or for the time it took to sublimate away all of a comet's ices, I could test different theories for the origin of comets. The key factor was whether the model could reproduce the distribution of orbital energies shown in Figure 5.2.

I was able to show that Oort's hypothesis of a distant comet cloud worked very well. By "tuning" the model input values, I was able to reproduce Figure 5.2, and in the process learn a great deal about the physical processes affecting comets. The model results showed that a typical comet from the Oort cloud passes through the planetary system only about five times. They also showed that about two-thirds of all the long-period comets are ejected to interstellar space. The remainder are lost to a variety of processes including random disruption or disintegration, loss of all volatile ices, and impacts with the Sun. And a very few comets, less than one in a thousand, remain in the solar system and evolve down to short-period orbits. We believe that Comet Halley is one such short-period comet whose orbit evolved inward from the Oort cloud.

An important feature of my computer simulation model was that it allowed me to test competing hypotheses for the origin of comets. A British astronomer named Raymond Lyttleton had proposed in 1948 that comets formed as the Sun passed through interstellar clouds, which caused dust and gas to be gravitationally focused in the Sun's wake. According to Lyttleton's idea, the dust and gas would not condense into a single solid nucleus as Whipple's theory suggested, but would just be a gravitationally bound swarm of material, called a "sand bank." Lyttleton estimated that the focus point for forming the sand banks would be only about 150 AU behind the Sun in its path, and thus comets would appear to come from much closer than Oort's proposed comet cloud. My computer model showed that Lyttleton's idea did not work. A focus point at 150 AU, or anywhere close to that, did not give the right distribution of orbital energies.

The Halley Armada

So after four years of work on it, I defended my thesis and graduated from UCLA. About the same time as George Abell was questioning me in 1977 about the approach of Comet Halley, scientists around the world were trying to decide what to do about it. Since Halley was such a large bright comet, it would provide a literally "once-in-a-lifetime" opportunity to study a very active comet in detail. For the first time ever, it also provided an opportunity to send spacecraft to encounter the comet.

To support myself while I was a graduate student, I had taken a part-time job at the nearby Jet Propulsion Laboratory (JPL) in Pasadena, CA. JPL is NASA's lead center for planetary exploration. JPL built the first United States satellite, Explorer 1, as well as the very successful Mariner, Viking, and Voyager spacecraft. My job at JPL involved assisting engineers in designing spacecraft trajectories to the planets and other targets. I worked in the Advanced Projects Group, a very talented and innovative bunch of trajectory specialists who's motto was "Any body, any time."

I was lucky enough to be assigned to help in the planning effort for a Halley mission. Initially, the engineers proposed a very ambitious mission that would match orbits with the comet and then go into orbit around the icy nucleus. Remember, Comet Halley goes around the Sun the opposite way from the Earth's motion, so it would take some pretty involved maneuvering to get the spacecraft to turn its orbit around. My particular job was to make computer-based movies of what the trajectory would look like in three dimensions, so it could be demonstrated to NASA managers and other technical audiences. These movies were primitive in comparison with current computer-generated graphics, but at the time they were the state-of-the-art (for many years one of our movies was used at Disneyland as part of the Ride to Mars). It was a lot of fun developing the special software that gave the best possible view of the spacecraft's convoluted path to rendezvous with Comet Halley.

At the same time, I was asked to work on theoretical modeling of the nucleus of Comet Halley so that the engineers at JPL would better understand the harsh environment that the spacecraft would encounter at the comet. I had been talking with one of my former professors at UCLA, Hugh Kieffer, about using a special computer program he had developed for estimating the temperature on the surface of Mars, and adapting it to the problem of predicting the surface temperatures and gas production rates from cometary nuclei. This was an exciting piece of new science. We were able eventually to learn quite a bit about the nucleus of Comet Halley by comparing our model predictions with both spacecraft and ground-based observations.

Sadly, at this point I also started learning about the realities of the space program. Halley was not the only body in the solar system. There were many other design studies underway at that time to send spacecraft to various planets. These included an orbiter and atmospheric entry probe for Jupiter, a radar mapping spacecraft for Venus, and a spacecraft to fly out of the ecliptic and pass over the Sun's north and south poles. The scientists and engineers who were designing those missions were just as excited about their targets as my comet colleagues and I were about Halley. In addition, the federal budget deficit was rising and the NASA budget was suffering severe cost-overruns from the construction of the Space Shuttle. As a result there was considerable pressure not to start any new missions.

And so, the plan for a Halley rendezvous mission never got off the drawing board. In its place, NASA and the European Space Agency (ESA) proposed a joint mission which could be launched later, allowing for better planning of the budget expenditures and time to develop several key technologies required by the mission. The new mission would rendezvous with a smaller, less active short-period comet called Tempel 2. A special feature of the mission was that it would still fly by Comet Halley, and would release a smaller spacecraft, built by ESA, that would fly very close to Halley's nucleus and make crucial observations and *in situ* measurements.

But this mission too was doomed. In 1979 further Shuttle cost-overruns caused NASA to cut a key new technology development pro-ram, solar electric propulsion (SEP), in order to reduce the overall budget. Without SEP, the Halley–Tempel 2 mission could not be flown.

This was a tremendous disappointment for all of us who had hoped we might finally get to see a cometary nucleus up close. We had all seen how our knowledge of the planets and their satellites had experienced a quantum leap forward with each new spacecraft mission, and we desperately wanted that same advance in our understanding of comets. Also, many of us strongly believed in the correctness of Whipple's icy conglomerate hypothesis, and a close-up photograph was the proof we needed to convince everyone else.

To the rescue rode our European colleagues at ESA. The ESA scientists and engineers decided that they would still build their Halley encountering spacecraft, and would launch it themselves on their new Ariane rocket. This was a bold move for ESA, which had never launched an interplanetary mission to a specific target before. But the commitment to do it was there, and the ESA scientists and engineers made it happen. They called their spacecraft Giotto, after the fourteenth century Renaissance painter who had created a fresco showing a comet hovering over the manger in Bethlehem containing the baby Jesus.

Other countries also joined the Halley armada. The Soviet Union proposed taking two Venus spacecraft that it was building, and targeting them both to fly by Halley after they went to Venus. The Soviet missions were called Vega 1 and 2, "Vega" being a contraction of "Venus-Halley" (unfortunately there is no "h" in the Russian alphabet so "Ve-ga" was the closest they could get). And the fledgling Japanese space program also decided to send its first two interplanetary spacecraft to Halley. These were called Sakigake ("Pioneer") and Suisei ("Comet").

Some US scientists were involved in the ESA mission, either helping to build the scientific instruments or participating in the data reduction following the flybys. One US investigator, John Simpson, at the University of Chicago, even managed to sneak his instrument onto the

Soviet missions. This was the era of "perestroika" and "glasnost," and relations with the Soviets were definitely improving.

Through a colleague I sent a message to the head of the Soviet program, Roaald Sagdeev, asking if I could participate in the imaging experiment on the Vega spacecraft. Along with the request I sent a present. It was a photograph I had taken of Sagdeev at a scientific conference in Austria in 1984. During a moment of levity on a week-end sightseeing excursion for the scientists, Sagdeev has been invited to lead a local town band. My photograph showed him doing this, using a wine bottle as a baton. For whatever reason, Sagdeev invited me to Moscow for the encounters. I was ecstatic! My bosses at JPL were too and bought me a new, state-of-the-art computer to do the image processing and analysis (called the "Comet Minicomputer," the machine was quickly nicknamed "Comi"). But NASA officials were still very nervous over the increased level of American–Soviet contacts and they refused to approve my travel to the Soviet Union. No amount of protest would change their decision. From the heights of ecstasy, I plunged into the depths of depression. I was left to gaze at Halley from afar.

Still, there was plenty to do. Following graduation, my part-time job at JPL had turned into a permanent one, and I was working on the Jupiter orbiter mission, called Galileo. At the same time, Kieffer and I were working on our comet thermal models and published several papers predicting the gas production rates expected from Halley. By comparing our predictions with spacecraft and telescopic observations, we were able to estimate some of the physical properties of the icy-conglomerate material on the surface of the cometary nucleus. In addition, I was busy with my computer-based dynamical models of the Oort cloud.

Yet somehow, Halley and I kept crossing paths. One of the scientists on our Galileo team, Fred Taylor, was a professor at Oxford University in England. While I was on a working visit there, Fred and his wife offered us a tour of the town. At one point we passed by a house with a plaque stating that it had been Edmond Halley's home when he had been a professor at Oxford, about 260 years earlier. For a cometary astronomer like myself, this was like a pilgrimage to Lourdes! Just around the corner was a very popular pub, the Turf, where we stopped for lunch. As we sat in the garden eating our meal, it suddenly occurred to me that Halley may have visited this same pub and eaten in this same garden, washing his dinner down with a tasty English lager or his beloved brandy. I checked with the barmaid and, sure enough, the Turf had been operating on that spot since the early sixteenth century. Scientists are not supposed to believe in ghosts, but on that spring afternoon it would not have been difficult to convince me that Halley's spirit might be hanging around that pub garden.

Back in Pasadena, new research topics kept cropping up. In 1980 Luis Alvarez and colleagues had published their breakthrough paper showing that the Cretaceous–Tertiary extinction 65 million years ago, the event that killed off the dinosaurs and 90% of all the life on Earth at that time, was caused by the impact of a 10-km diameter comet or asteroid on the Earth. This was a radically new idea and many geologists and other scientists refused to believe it. I was quickly drawn into this controversy, estimating the rates at which long- and short-period comets impacted the Earth. It turned out that the impactor flux is dominated by asteroids, but comets still make up about 10% of the large objects striking the Earth.

In 1980 astronomer Jack Hills, of Los Alamos National Laboratory, spent the summer working with us at JPL and suggested an interesting new idea. Jack proposed that on rare occasions, about once every 500 million years, a star will actually pass right through the Oort cloud, only about 3,000 AU from the Sun. This might have a catastrophic effect. Many comets would be ejected to interstellar space but a large number would also be thrown into the planetary region. These so-called "comet showers' would literally flood the planetary zone with comets, raising the impact rates of comets on the Earth by a factor of 300! As a result, there could be a dramatic increase in the probability of a biological extinction event. We actually think that there is evidence of a modest cometary shower about 36–38 million years ago, during a geological era known as the late Eocene. Several large craters and other evidence of impacts on the Earth have been found and dated to this period, which is also identified with a modest biological extinction event.

Dirty snowballs and icy rubble piles

My interest in comet showers led me to a collaboration with another astronomer, Piet Hut, and in 1985 he invited me to spend several months working with him at the Institute for Advanced Study in Princeton, New Jersey. As we worked on understanding the details of comet showers, I also had time to think about several other scientific problems. At this time Comet Halley was racing towards the Sun, and public interest in the comet was growing rapidly. I made several trips for the American Astronomical Society, giving lectures around the country to standing-room-only audiences, even appearing on a local TV show with a name something like "Good Morning Cleveland" before one such lecture in Ohio. At another lecture, a little 95-year-old lady who had seen Halley's Comet in 1910 came up to me after my talk. "It wasn't like your slides," she exclaimed. "It was all over the sky!" Things were getting very exciting.

But one question kept nagging at my mind. What is the nucleus of Comet Halley going to look like? The spacecraft flybys were planned for March, 1986, only a few months away. What would the cameras find? Because of Whipple's "dirty snowball" model, we all tended to imagine a fairly compact object, probably irregular in shape, but still very much like a snowball you would make on a cold winter's day. Would that really be the case?

There were hints that comets might look different. For example, comets are sometimes observed to split, that is, they break into several separate pieces. These pieces travel together in the same orbit around the Sun, slowly drifting away from one another. After a few days or weeks, the smaller fragments fade away, and only the main comet nucleus remains. A spectacular example of this was a short-period comet called Biela which was seen to split in 1846, returned as a double comet in 1853, and then was never seen again. The comet apparently disintegrated completely. So it seemed that comets were very fragile objects, and small pieces were often breaking off.

A second important fact came from theoretical studies (by various scientists) of the formation of comets and planets. As small bodies like asteroids and comets came together and grew into planets in the primordial solar nebula, they released energy, known as gravitational potential energy. Gravitational potential is what holds you on the surface of the Earth, and considerable energy must be expended to lift you into outer space—that's why rockets are so big and powerful. In the case of the forming planets, the gravitational potential energy was released as heat, and that heat melted the planets, allowing them to form into spheres. The more massive the planet, the greater the gravitational potential energy that was released in assembling it.

But comets are too small to release much gravitational potential energy. The ices and dust grains in them have never been melted or modified. It is this "primitive" composition that makes comets so important to understanding the origin of the solar system. Frozen in the comets is a chemical record of the composition of the solar nebula 4.5 billion years ago. Knowing that starting point for making the solar system would be an invaluable piece of information for understanding how the planets formed. As a result, many of the science instruments on the Halley flyby spacecraft were designed to measure the composition of the comet's outflowing dust and gas.

For me, the very low gravitational potential energy of the comets meant that the nucleus would also preserve the *structure* they had at the time the dust and ice grains came together billions of years ago. When we make a snowball, we compress the snow with our hands, making it into a single, monolithic structure. Imagine what a snowball would look like if we stuck a lot of tiny snowballs together, but did not compress them into a single ball. Could that be what a comet nucleus would look like?

I calculated the gravitational potential energy for a typical cometary nucleus and, sure enough, it was not enough to melt the comet. It was not even enough to raise the temperature of the nucleus by one degree centigrade! I also calculated the speed at which the pieces of the comet would come together and it was only about 1 to 3 meters per second, or about 3 to 10 feet per second, something like the speed of a brisk walk. This speed might result in some modest crushing as the fluffy snowballs came together, but it could not cause much significant heating.

Another piece of evidence also pointed to comets being loosely bound collections of smaller dirty snowballs. Comets would occasionally experience outbursts, sudden flares in their brightness where large amounts of dust and gas were released. These could be explained if an old snowball on the surface of the nucleus broke away, exposing fresh ices below it.

The three months I spent at the Institute for Advanced Study gave me time to think about this problem. Without the constant interruptions of my telephone at JPL and the work I was doing supporting spacecraft missions, I had the time to peruse the scientific literature and to carefully work out all the details of my new idea. By Thanksgiving I was ready to write it all down. I spent that Thanksgiving at my brother's house in Boston and, between eating turkey and playing with my nieces and nephews, I used the time to write a paper for the scientific journal *Nature*.

But what should I call this model? Researchers studying asteroids had already developed what they called "rubble pile" models. They had suggested that asteroids in the asteroid belt were blasted into rubble when they collided with one another. But then the rubble would gravitationally come back together into one or more bodies, a pile of rubble so to speak. What I was describing was very similar, except the "rubble" was not the broken up remains of earlier comets, but rather primordial condensations from the solar nebula. And so I called the idea the "primordial rubble pile."

A common phenomenon in science is that several researchers may propose the same solution to a problem almost simultaneously. This is a natural result when scientists work in the same area, using the same data, trying to answer the same questions. So just about the same time I was proposing my primordial rubble pile model, two colleagues, Bertram Donn (NASA Goddard Space Flight Center) and David Hughes (University of Sheffield, England), proposed a very similar model which they called the "fluffy aggregate." Donn's and Hughes' arguments were similar, pointing to the random accretion of microscopic dust and ice grains in the solar nebula, and how that process might be mimicked at larger sizes as icy-conglomerate snowballs collided and stuck. David Hughes was a reviewer on the paper I submitted to *Nature*, so we enjoyed some very useful exchanges on the topic.

My paper was published in *Nature* in March, 1986, the same week that Giotto flew within 600 km of the Halley nucleus. Sitting in a conference room at JPL with other scientists, I anxiously awaited the pictures that might confirm my hypothesis. A week before, the two Vega spacecraft had flown by Halley but much farther away than Giotto's planned encounter. Their pictures had shown an irregularly shaped nucleus, but were too far away to reveal any surface details.

My emotions on that spring afternoon were a bizarre mixture of excitement, anxiety, and depression. I was excited because I was finally going to see a comet nucleus up close and personal, and my "primordial rubble pile" model might be proven right. I was anxious because I might just as easily be proven wrong. And I was depressed because I was still smarting from NASA's denial of permission for my travel to Moscow for the Vega encounters.

As we waited, Ray Lyttleton, the creator of the competing sand bank model, walked into the room. Lyttleton often spent time at JPL working with an old friend on various research problems. In a few minutes he was going to have his sand bank model proven wrong. Lyttleton glanced nervously at the TV monitor for a minute or two and then walked away.

As the pictures came in, we strained to see details. Unfortunately, the raw images were difficult to decipher. It sometimes requires considerable computer processing to bring out the details in spacecraft images. In addition, someone at the European receiving center had thought to add false color to the black and white images, and the color only served to further confuse the appearance of the images. Perhaps the pictures would make better sense when the spacecraft got really close to the nucleus.

And then the pictures stopped. Some 14 seconds before closest approach the tracking antennas on Earth lost contact with Giotto. It turned out that a large cometary dust grain, perhaps the size of a dime, had struck the spacecraft. At the flyby velocity of 68.4 km per second (152,000 miles per hour!) that little pebble carried the momentum of a cannon ball. It caused the spinning spacecraft to wobble about its axis, and pointed the spacecraft's high gain antenna away from the Earth. We had always known that this was a possibility, but we all had hoped that Giotto would beat the odds. Within a few tens of minutes automatic systems onboard the spacecraft regained control, but by then the damage had been done. The closest approach data were lost, along with what should have been the best pictures of the nucleus surface.

Still, the images obtained up until that point gave us a tantalizing view of the nucleus of Comet Halley. After a few weeks of computer processing, a clearer picture of the nucleus began to emerge (see Figure 5.3, color section). What we saw was a very dark, irregular body, sort of potato or peanut shaped, with all sorts of bumps and protrusions. The picture was not quite sharp enough to see the underlying nucleus structure, but it strongly hinted at a rubble pile structure.

Seeing that picture was incredibly exciting for me, and not only because it suggested that my ideas about a rubble pile structure were correct. For me, that picture was the Holy Grail of solar system science. I was looking at a picture of the creation of the solar system, a picture of one of the earliest bodies to form in the solar nebula. The scientists on Giotto had captured a vision of Genesis and it was an awe inspiring image to behold.

Later comets would provide further evidence that cometary nuclei were indeed rubble piles. Most notable of these was Comet Shoemaker-Levy 9 which had been captured into orbit around Jupiter. That comet was discovered in 1993 after it made a very close approach to Jupiter, during which the planet's gravity tore it apart into its component pieces, and then self-gravity reassembled the pieces into a string of 21 separate comets. Two planetary scientists, Erik Asphaug (of NASA's Ames Research Center) and Willy Benz (of the University of Arizona), showed that the breakup and reassembly of the nucleus could only be understood if it was a weakly bound, rubble pile structure. Later when the fragments crashed into the atmosphere of Jupiter in July 1994, resulting in incredibly huge explosions, physicists Mark Boslough and David Crawford (both at Sandia National Laboratory) showed that some aspects of the explosions could best be explained if the reassembled fragments were also rubble piles.

These confirming proofs were still to come when I flew from Los

Angeles to Rio de Janeiro a few weeks after the Halley spacecraft encounters in March, 1986. I was enjoying all the hubbub that Halley had generated, and now I was going to the southern hemisphere to see the comet for myself. Good fortune had smiled on me and I had been invited to lecture onboard a luxury cruise ship as it sailed from Rio to Fort Lauderdale, Florida. Bill Kaula had been wrong. You could send a cruise there!

As I gazed at the comet and the spectacularly beautiful southern sky, I could not help thinking about George Abell and his marvelous enthusiasm for astronomy. Here before us was Halley's Comet, a celestial messenger from the past that held the secrets of our origins. It was the same comet that humankind had gazed at for two millennia or more, a regular visitor waiting to tell us all it knew. And now for the first time we were learning how to listen and learning how to unravel the message locked in the comet. George would have loved it.

Figure 5.3

Composite image of the nucleus of Comet Halley, as photographed by the Giotto spacecraft on March 14, 1986. The Sun is shining on the nucleus from the left. The nucleus is seen to be a dark, highly irregular body, about 16 km in length and about 8 km wide. Bright jets of dust and gas are being emitted from active areas on the sunlit hemisphere of the nucleus. The image was created by putting together many pictures of the nucleus taken as the spacecraft raced towards it at 70 km per second. The Giotto camera was aimed at the upper left hand corner of the nucleus, so the detail is best in that region of the composite picture. Image courtesy of H. U. Keller, Max-Planck-Institut für Aeronomie.

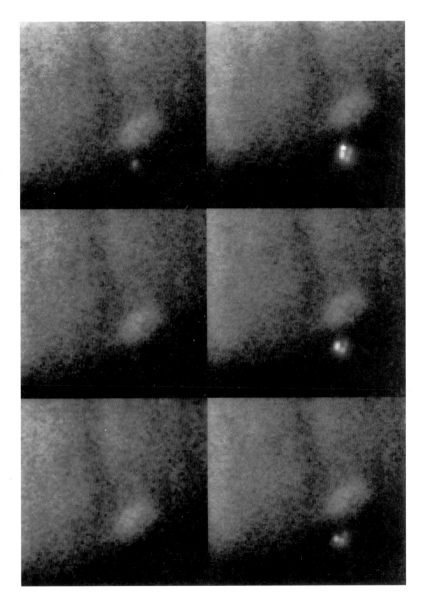

Figure 6.1

Io's volcanos emerge from behind Jupiter, in this time sequence of infrared images taken over a four-minute interval in August 1995 at the NASA Infrared Telescope on Mauna Kea. Io is in Jupiter's shadow, so its disk is invisible, and only the volcanos can be seen. Six separate volcanos, which appear from behind Jupiter in turn, can be distinguished in the original data. The colors have been added later, but approximate the colors that we would see if our eyes were sensitive to infrared light. The brighter patch visible on Jupiter is the Great Red Spot. Photograph by John Spencer.

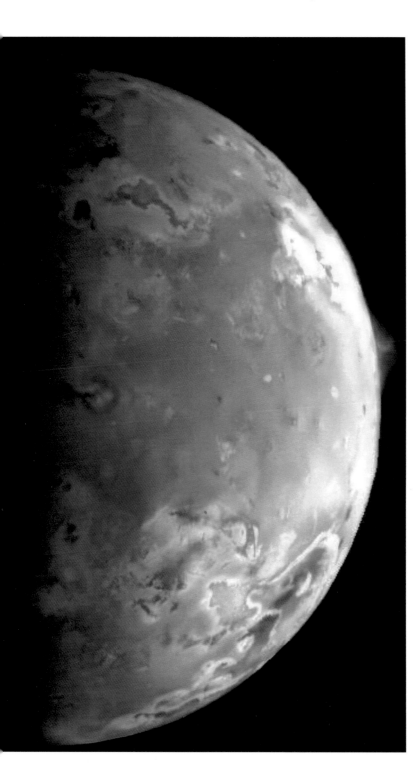

Figure 6.2

The Galileo spacecraft took this dramatic image of Io in June 1997. On the edge of the disk a new plume, Pillan, rises 120 km above the surface against the blackness of space, while near the terminator (the line dividing the dayside and nightside), the volcanic plume Prometheus casts a long shadow over the surface. The Pillan plume was first seen in Hubble Space Telescope images taken a few days after this image. Multicolored patterns on the surface are deposits of sulfur and sulfur dioxide spewed out by the volcanos. Photograph from NASA.

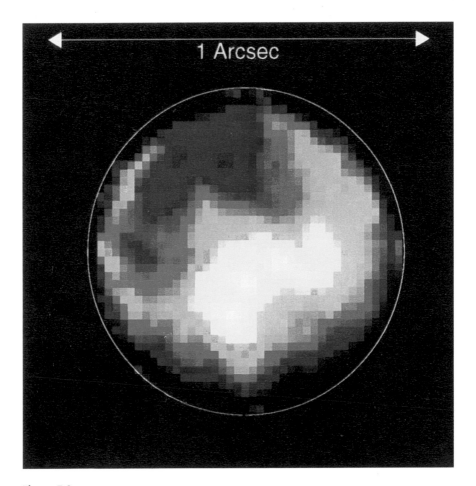

Figure 7.2

Titan seen through a methane window from the 10-meter Keck Telescope, using a technique called "speckle interferometry" to compensate for the Earth's atmospheric turbulence. Image constructed by McIntosh and collegues, and reproduced courtesy of Lawrence Livermore Laboratories.

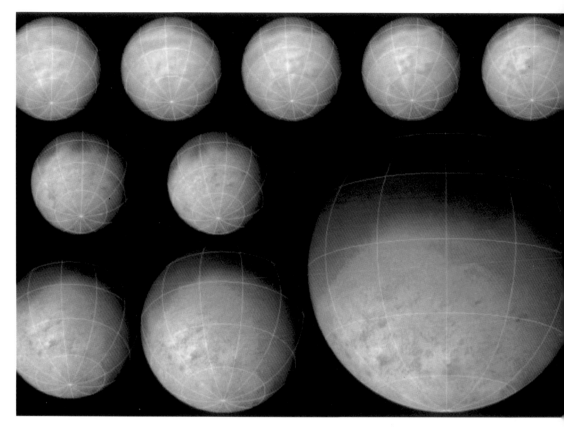

Figure 8.4

Triton approach sequence, overlaid with a latitude-longitude grid. Details on Triton's surface unfold dramatically as the resolution changes from about 60 km per pixel at a distance of 5 million km for the image in the upper left to about 5 km per pixel at a distance of 0.5 million km for the image in the lower right. Mainly looking at the southern hemisphere, we see Triton rotate retrograde (counterclockwise) over an observation period of 4.3 days. Image courtesy of the NASA Planetary Data System Photojournal.

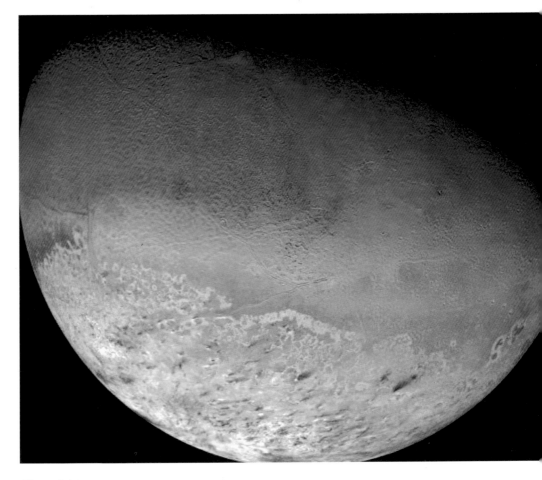

Figure 8.6

Computer photomosaic of the best Triton images. The equator runs approximately through the center of the bright swath across the middle of the mosaic (compare with Figure 8.4). The north pole was in darkness at the time of the Voyager encounter. Triton's surface is covered with deposits of solid nitrogen with small admixtures of methane. The bluish tinge is characteristic of fresh frosts, while the reddish tint is thought to be due to partially irradiated methane. Image courtesy of the NASA Planetary Data System Photojournal.

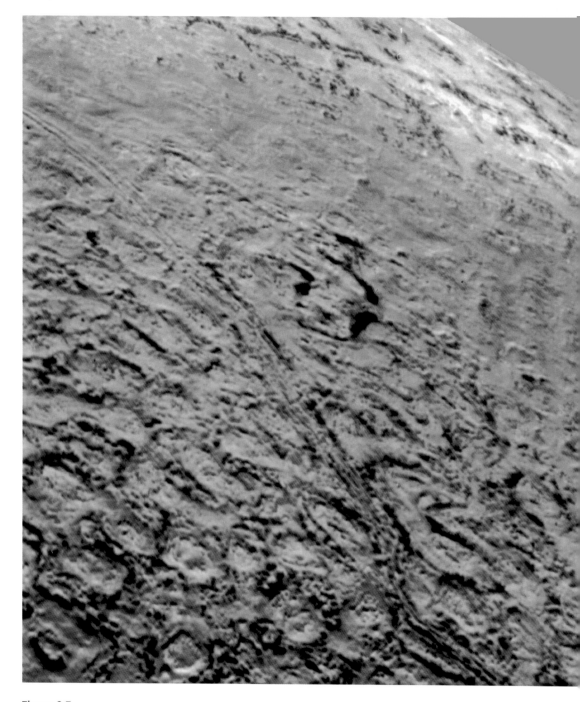

Figure 8.7

Cantaloupe terrain at the bottom and polar terrain at the top, in this high-resolution Voyager image taken from a distance of 40,000 kilometers. Each cantaloupe "dimple" is about 25 to 35 kilometers across. A tectonic fissure runs through the cantaloupe terrain, probably formed by the extension of Triton's icy crust. Image courtesy of NASA JPL.

6

Io and I

JOHN SPENCER

John Spencer. England, where John Spencer grew up, with no real space science program, and cloudy, north Atlantic skies, is a long way from the planets. Nevertheless, John was bitten by the planetary research bug, and siezed that as an opportunity to come to America. And for most of the almost 20 years since he did that, John has invested much of his incalculable energy into studying the large Galilean satellites of Jupiter. Of these four worlds (Io, Europa, Ganymede, and Callisto), John's special fascination is in the bizzare, Dante-esque Io (pronounced, EE-oh by some, and Eye-oh by others). As John describes to us, his deep fascination with Io began the day the first Voyager spacecraft reconnoitered this world, and continued through graduate school in Tucson, through his postdoctoral fellowship in Hawaii, and through his tenure at Lowell Observatory, in Flagstaff, Arizona. John and his wife, Jane, actually live on Mars Hill, just meters away from his office at the Observatory. In the essay below, John takes us on a fascinating journey of personal exploration of his favorite of Sol's worlds, Io.[1]

In the Spring of 1970, in Lancashire, England, a teenage boy (me) drags his indulgent visiting relatives out of the warm house into the back garden to show them something wonderful through his borrowed 3″ telescope: the bright sparks of Jupiter's four big moons, the Galilean satellites, arrayed on either side of the giant planet. In the spring of 1995, on the summit of Mauna Kea, Hawaii, a planetary astronomer (yes, me again) trains the 3-meter NASA Infrared Telescope on the strangest of those moons, volcanic Io, and sees an enormous volcanic eruption, bigger than any he has seen before, its brilliant infrared glow

---

[1] This chapter is a greatly expanded, and updated, version of an article that first appeared in *The Planetary Report*, the magazine of the Planetary Society, in July 1997.

overwhelming Io's disk. This time, his audience is a bit larger, for thousands of schoolchildren in North America and Europe, participants in the annual "Jason Project," are listening in, and hear about the discovery as it happens. But in each case, the sense of wonder inspired by the awesome Jupiter system, and the thrill of sharing that wonder with others, is the same.

Why do I find the Jupiter system so fascinating? Perhaps because it is the place where, as we move away from the Earth, things start to become seriously strange. Jupiter is near enough that it dominates our skies, and has been familiar to us for as long as we have been human, yet far enough away that the terrestrial rules no longer apply. Here's a planet with no surface, where multicolored storms rage for decades, places where the rock is made of ice, where the metal is made of hydrogen, where the snow is made of sulfur dioxide, where moons come by the dozen, where the skies are full of strange glows and invisible, deadly, blizzards of high-energy atoms, where worlds are melted and comets torn apart by the force of gravity. All arrayed for our viewing pleasure, a mere half-billion miles away.

Most fascinating of all the worlds in the Jovian system, at least for me, is bizarre Io, home of the only known active volcanism beyond Earth. Studded with rugged mountains, pitted by vast volcanic craters, buried in huge outpourings of lava, Io's surface looks like nothing else in the solar system. The volcanos have covered the surface in colorful, volatile, compounds of sulfur and oxygen, some of which evaporate to form a tenuous, malodorous, atmosphere. A ton of oxygen and sulfur is torn away from Io every second by an intense bombardment of electrically charged (ionized) atoms, caught in Jupiter's rotating magnetic field, that slam into Io at a speed of 50 kilometers per second. The atoms that are removed are themselves ionized and are swept around Jupiter by the magnetic field, some returning to rip away more of Io in turn. The rest of the atoms, flung out by centrifugal force, spread throughout Jupiter's magnetosphere (the region of space dominated by Jupiter's magnetic field) in vast, glowing, energetic clouds, staining the surfaces of the other moons, and generating intense auroral displays at Jupiter's poles.

A boy in England

Back in 1970, when I was first getting to know the Jupiter system in my back garden, very little of this story was known. Powerful telescopes existed, but the instruments available to analyze their light were primitive, and Io was so small and far away that it appeared as little more than a blurry dot in even the best telescopes. Jupiter was enjoying the last couple of years of its 4.5 billion year era of privacy before the first spacecraft arrived from Earth. Scientists in the young field of planetary science were focusing their attention closer to home, where the Apollo astronauts were conducting their personal surveys of our own, far more sedate, Moon. Still, though it hadn't made it into the astronomy books that I was devouring at the time, people were beginning to realize that Io was very peculiar.

Io's strong yellow color was different from anything else in the solar system, and unlike Jupiter's other big moons, Europa, Ganymede, and Callisto, measurements of its infrared spectrum showed that Io had no water ice on its surface. When Io emerged from the cold of Jupiter's shadow, its surface seemed to be unusually bright for a few minutes, as though frost had condensed on its surface during the eclipse, and was being burned off by the returned Sun (but without water, what kind of frost could that be?). Most strangely, Io was somehow controlling part of the cacophony of radio noise that Jupiter broadcast into the solar system: the noise rose and fell in time with Io's orbit around the planet. As a result of these findings, Io began to get the attention of astronomers, and more telescopes were pointed in its direction.

Meanwhile, I was working my way through grammar school, encouraged in my astronomical and academic interests by some great schoolteachers and by my parents—my Dad searched long and hard for a telescope for me, until he persuaded some friends to lend us that 3″ refractor. Though I was only dimly aware of the developments, Io was revealing more and more peculiarities to those with larger telescopes than mine. In 1974, astronomer Bob Brown of Harvard discovered a glowing cloud of sodium gas surrounding Io, and this was soon followed by the discovery of the more massive but less conspicuous clouds of sulfur and oxygen. The first spacecraft to penetrate Jupiter airspace, Pioneer 10, discovered that Io had a tenuous ionosphere, an envelope of ionized atoms. Io's peculiar color seemed to be due to vast amounts of sulfur on the surface, but there were hints of other, still unidentified, materials too.

Furthermore, something was very odd about Io's infrared radiation. All objects that absorb sunlight, and are warmed by it, re-radiate that energy as infrared light, and astronomers have long known how to measure infrared radiation to determine the temperatures of bodies in the solar system and beyond. Io didn't fit the usual pattern. Dale Cruikshank and David Morrison, at the University of Hawaii, looking at an infrared wavelength of 20 microns, came to quite different conclusions about Io's surface temperature than did Caltech's Olaf Hansen, working at 10 microns, and it wasn't clear how to reconcile the two observations. Then, in early 1978, Fred Witteborn of NASA Ames Research Center saw a brief, intense, burst of infrared light from Io which seemed to show that, incredible as it seemed, part of the surface had suddenly become hundreds of degrees hotter than its surroundings.

Io was baffling, but help was on the way. In 1977, NASA launched what would become the most spectacular missions in the young history of planetary exploration. Voyager 1 and Voyager 2, spacecraft far more sophisticated than Pioneer 10, were dispatched towards Jupiter with orders to return thousands of detailed images of Jupiter, its moons, and the planets beyond. And not just images, but infrared and ultraviolet spectra, and detailed measurements of plasmas and magnetic fields. Voyager promised, and then delivered, a revolution in our understanding of the wonders of the outer solar system.

By the time of the first Voyager launch, I had convinced myself that geology, with its practical applications, was a more realistic career choice than astronomy, and I was spending the summer in a caravan (house trailer to Americans) in the Scottish Highlands, learning how to make geological maps. I had one year to go before I graduated from Cambridge University with my geology degree. I saw a job with an oil company in my future, and wasn't sure how I felt about the prospect. But despite my temporary confusion about my direction in life, and my ignorance of just how peculiar Io and the Jupiter system were turning out to be, I knew enough to be thrilled that the Voyagers were on their way to Jupiter. I could hardly wait till March 1979, when Voyager 1 would arrive.

To the States

By the time I graduated, in June 1978, I had decided against the oil companies (they hadn't seemed too keen on hiring me anyway), and had accepted a place in a PhD program in geology at the University of Hull, in Yorkshire, to start in September. But I had special plans for the intervening summer. On June 17 I flew to Houston, Texas, to take a job as a summer intern at the Lunar and Planetary Institute (LPI), next to NASA's Johnson Space Center. It was a summer that changed my life. Working with effervescent LPI planetary geologist Peter Schultz, my project involved classifying and counting craters on Apollo pictures of the lunar farside. Though much of the actual work was tedious (those craters started to look pretty similar after the first few hundred), I was thrilled to be working on Apollo data, and to be surrounded by the trappings of the space program. After work, I would spend hours browsing through the files of photos brought back by the Apollo astronauts, or the new Mars images from the Viking probes, touching a world that I had never imagined having access to. Maybe an astronomical career wasn't so unrealistic a goal, after all.

I returned to England and the University of Hull in a daze—it was hard to come back to earth and concentrate once more on terrestrial geology, though I did my best. As 1978 turned into 1979, a new distraction loomed: Voyager 1 was now bearing down on Jupiter, heading for its early March flyby. It was a frustrating time for me: coverage of the mission in the British media was skimpy, and in those pre-World-Wide-Web days, there was no way for me to access the latest Voyager news for myself. The few glimpses of the mission results that I could get—a picture in *Time* magazine of Io and Europa crossing the storm-tossed face of Jupiter, a movie of Jupiter's rotation on the TV news that even the newscaster called "stunning"—burned deep impressions on my brain that remain to this day. On the day of encounter, March 5, when Voyager 1 plunged headlong through Io's sulfur cloud and a mere 22,000 kilometers below the south pole of Io itself, I re-read Arthur C. Clarke's thrilling account, in *2001: A Space Odyssey*, of the fictional Jupiter flyby of the *Discovery* and her crew, and tried to imagine myself onboard Voyager, seeing what it saw.

But I was luckier than most information-starved Voyager fans. Two weeks after the flyby I was once more on a plane to Houston, to attend the annual Lunar and Planetary Science Conference—Pete Schultz would be presenting the results of our previous summer's work at the meeting. Finally, I would get all the Voyager details! I can still clearly remember the moment when I first looked over the display of Voyager results that had been set up at the conference. Incredible images of Jupiter and all the Galilean moons were on display—a gallery of wonders that would have been unimaginable a few weeks earlier. One picture of Io was even more remarkable than the rest: a closeup of the edge of its disk showed what looked like a bright yellowish cloud rising off the surface. My mind raced—could we have caught the moment of a huge meteorite impact on Io's surface? The odds against that seemed overwhelming. I checked the caption—it wasn't an impact, but a volcanic eruption, currently in progress, one of eight that Voyager had discovered! I had been so conditioned by earlier spacecraft images that had revealed the Moon, Mars, and Mercury to be geologically dead that, like most planetary scientists, the idea of current geological activity on another world seemed incredible. But there was Io spouting off in volcanic exuberance, in defiance of our expectations.

Voyager's discovery of active volcanism on Io made sense of most of the list of peculiar things about Io that had been accumulating over the previous couple of decades. Everything fell into place. It was the volcanos that produced the excess heat that had made Io's infrared radiation so hard to understand. It was probably the volcanos that had boiled off the water, and had covered the surface in sulfur compounds, producing Io's peculiar color, and fed the sodium, sulfur, and oxygen clouds that surrounded Io, via the thin atmosphere. That atmosphere, generated by the volcanos, made the electrical connection through Jupiter's magnetosphere that generated the electric currents that made the radio signal that Io controlled.

More wonderful still, as I learned at that conference, even the presence of the volcanos in the first place made sense. In the kind of coup that most scientists only dream about, Stan Peale, Pat Cassen, and Ray Reynolds had published a paper in the journal *Science* just weeks before the Voyager discoveries that predicted massive heating of Io's interior due to distortions of its shape by Jupiter's gravity. "Widespread and recurrent surface volcanism" was a probable result, they said. Rarely in scientific research has such a bold and surprising prediction been borne out so quickly. The trio got a standing ovation when they presented their results at the conference.

What Peale, Cassen, and Reynolds had figured out was the consequence of a long-known and remarkable fact about the orbits of the Galilean satellites. The three inner moons, Io, Europa, and Ganymede, are locked into a beautifully simple dance by their gravitational influence on each other. Every time Ganymede, the most distant of the three, orbits Jupiter once, Europa goes around Jupiter twice, and Io, the innermost, goes around four times. The match is so perfect that the three moons can never line up on the same side of Jupiter (the way they did in the movie *2001: A Space Odyssey*). Because of this resonance, each moon is tugged rhythmically by the others, and their orbits are distorted, so their distances from Jupiter are not constant. This has tremendous consequences because Jupiter's gravity has distorted the shapes of the moons themselves, raising tidal bulges like those that our Moon raises on the Earth and its oceans, but much bigger—the side of Io that faces Jupiter bulges 10 km toward the giant planet. As each moon's distance to Jupiter varies around its non-circular orbit, the size of the tidal bulge varies, and the moon is continually flexed: the solid surface of Io goes up and down by perhaps 40 meters every day. The flexing generates frictional heat (called tidal heat) in the interior, to an extent that depends mostly on the closeness of the moon to Jupiter (the closer to Jupiter, the bigger the tidal bulge and the greater the heating). On Ganymede, the tidal heating is currently of little consequence. On Europa, the heating is sufficient to warm the interior, smooth out the surface, and perhaps maintain a subsurface ocean of liquid water. But Io, closest to Jupiter, suffers the most. Tidal heating pours so much energy into Io that the heat must escape through the surface at a rate, per unit area, *forty times* the average terrestrial rate. No wonder Io is hyperactive.

I returned from the Houston conference to England with a folder full of color copies of the Voyager images, and a head full of new ideas. The solar system was stranger and more wonderful than I could have imagined. Though I spent the summer of 1979 once more traversing the beautiful, lonely, wilds of the Scottish Highlands with a rock hammer, trying (unsuccessfully) to unravel a small part of the Earth's ancient past, those Voyager pictures were pasted to the wall of the caravan where I was living, and my mind was constantly wandering off-planet, out towards Jupiter. At the end of the summer, I abandoned my geology PhD program and sent in my applications to planetary science graduate schools in the United States. I wanted to be a planetary scientist.

Io time

And so, in mid-1980, I arrived in Tucson to start a new life as a PhD student in planetary sciences at the University of Arizona. I remember a moment of panic at the end of the long journey from London, while the plane dodged the August monsoon thunderheads on its descent into Tucson. I had given up on a PhD program back home, and now I was starting another one, far away in a place where I knew nobody. I couldn't face failing again. But I needn't have worried—planetary science caught my enthusiasm where terrestrial geology had not. When it came time to choose a dissertation topic, I would have liked to work on Io, but fascination with a subject is an inadequate foundation for a PhD thesis. I needed a "niche," something unique and substantial that I could contribute to the field and, like many graduate students, I had trouble finding that niche. All the great Io research topics that I could think of seemed to be undoable, or already being done by others. In the end, my dissertation work concentrated on Jupiter's other big moons: Europa, Ganymede, and Callisto. Still, I was often looking over my shoulder toward Io, and tried to keep up with the latest news from that corner of the Jupiter system.

In 1987, I finally earned my PhD, and started a postdoctoral fellowship at the University of Hawaii. There, I seized the opportunity to team up with veteran Io observer, Bill Sinton, who in 1979 had realized that volcanic heat was the reason that Io had seemed so peculiar in infrared telescopic observations, and that therefore Io's volcanos could be studied from Earth. Since then, Bill had been watching the infrared glow of the volcanos from the NASA Infrared Telescope Facility (IRTF) at Mauna Kea observatory on the Big Island of Hawaii, trying to understand how they behaved. We tried to continue his program, making a few improvements, but for a couple of years we had a frustrating time. It seemed that, whenever we went to the telescope, the clouds closed in. On the few occasions when the weather was clear, our equipment didn't work. We were also frustrated that, while we could measure the general volcanic glow from Io (weather and equipment permitting), it was difficult to locate and identify individual volcanos: Io was so small and far away, just a fuzzy dot in the sky, that usually we could only measure the total volcanic radiation. Only Loki, Io's largest volcano, could be distinguished on a regular basis, and that only by indirect techniques. It was hard to understand the volcanos when you couldn't get to know them as individuals.

We and the rest of the small community of Io volcano watchers were therefore looking forward to a period in early 1991 when, in a rare series of events, Europa would pass in front of (occult) Io once every few days for several months. By measuring the total infrared light from the blurry, merged, image of the two moons as the occultation progressed, we hoped to see sudden jumps in brightness as individual volcanos winked out or re-emerged from behind Europa. The timing of these events would tell us where the volcanos were, and the size of the jumps would tell us how big and hot they were. For a few months, we would have a clear picture of what was going on.

Getting the most out of these rare events would require careful planning. In June 1989, fellow "Iophile" Jay Goguen, a former Bill Sinton protege like myself, convened a small workshop in Pasadena to talk about the upcoming occultations. One of the participants was an amateur astronomer, John Westfall of the Association of Lunar and Planetary Observers, who was interested in watching the occultations in visible light, to refine our knowledge of the locations and orbits of the satellites. As we mulled over our plans, John pointed out that, while Europa occultations of Io happen only for a brief period every few years, Jupiter occults Io on every 1.8-day orbit of the moon, forever. Couldn't we locate and measure individual Io volcanos using Jupiter occultations, in addition to the Europa occultations? To our embarrassment, none of us highly trained professionals had thought of this. It seemed that it might be feasible: the fuzzy edge of Jupiter's deep atmosphere wouldn't slice across the volcanos as cleanly as the knife-edge of Europa, so the view wouldn't be as detailed, but the routine nature of the Jupiter occultations was a big advantage: we could look more frequently, over a much longer period. There was another, less obvious advantage: for part of every Jupiter occultation of Io, when Io passes behind the body of Jupiter itself, Io is also eclipsed by Jupiter, i.e., it is in Jupiter's shadow. In the darkness, Io's volcanos glow without competition from sunlight, and are much easier to see. The combination of a Jupiter eclipse and occultation was potentially a powerful one.

Also in 1989, the first digital infrared cameras were becoming widely available. These could take real infrared pictures of celestial objects, instead of just measuring the total amount of their radiation. We might expect that Io would be too small to show any details even with an infrared camera, but the first users of these cameras were reporting spectacularly sharp images from the telescopes on Mauna Kea, perhaps even sharp enough to see details on Io's tiny disk. Could we combine John Westfall's idea with this new technology, and use an infrared camera to take pictures of Io that could resolve and locate individual volcanos, refining their positions further by timing their occultations by Jupiter? It seemed worth a try, so Bill Sinton and I put in a proposal to the NASA infrared telescope to use their new infrared camera (called, with eccentric capitalization, ProtoCAM) to look at Io in early 1990.

We got results sooner than we expected. In December 1989, the team developing the new camera, led by IRTF staff scientist Mark Shure, were preparing for a test run of the camera on the NASA infrared telescope, and Jay Goguen and I independently suggested to them that Io would make a good test target. On the third night of the tests, Io was due to pass behind Jupiter, so we could also try the occultation technique. I decided to fly over to the Big Island, for the third night only, to help out with the Jupiter occultation. On the first evening of the tests, I was just going to sleep at home in Honolulu when the phone rang. An excited Mark Shure was on the line from Mauna Kea. "We can see Loki!", he exclaimed. Mark and his team were lucky enough to have caught the volcano during a bright eruption, and the images were indeed sharp enough to see it directly, as a bright point of light on Io's disk.

So it was with great anticipation that I flew to the Big Island two evenings later, the night of the Jupiter occultation. The details of that night are still very clear in my mind. It was two days before Christmas, and Hilo airport was in a festive mood, full of people getting together for the holidays, and greeting each other with leis. I left the bustle behind and drove past suburban houses garish with Christmas lights, then continued upward on the familiar mountain road, into the darkening rain forests and open ranchlands that ring Mauna Kea. I pulled off the side of the road at about 8,000 feet to admire Venus and Mercury, gorgeous in the western sky, and ran over something sharp in the process—I paid for my sightseeing with a flat tire. After a frustrating delay, I was able to get a lift up to the 13,500 foot summit of the mountain, where Mark and his team were already at work testing the camera. Finally, we turned the telescope to Io, and there it was: a crisply resolved disk, with Loki blazing forth in the north-east quadrant, even better than two nights earlier. In the rock-steady air of a perfect Mauna Kea night, we were seeing Io more clearly than anyone had since Voyager. Right on schedule, Io slid into Jupiter's shadow. Its disk disappeared into darkness, but Loki remained, a brilliant starlike point of light, glowing in the dark. Then, as Io continued on its orbit and passed behind the disk of Jupiter itself, Loki's glow held steady until it swept behind the planet and suddenly winked out. The time of the disappearance, we figured out later, precisely confirmed that the bright volcano was Loki. John Westfall's occultation idea had worked! We drove down the mountain in the dawn light of Christmas Eve exhilarated. It had been one of those rare and magical nights at the telescope when we had seen something that no one had ever seen before. We had watched a volcano set behind a planet.

But more was to come. Back in Honolulu, after the New Year, I spent most of a frustrating Saturday afternoon struggling with a research project that didn't seem to be going anywhere. Looking for some relief, I decided to take a closer look at those images of Io's occultation by Jupiter (Figure 6.1, color section). After running through the sequence of frames a few times, I thought I saw an extra point of light next to Loki, almost lost in the big volcano's glare, but disappearing behind Jupiter before Loki did. I gradually convinced myself that it was real, and called a couple of other people into the office to get a second opinion. They agreed that yes, there was something real there. It was a new volcano, never before seen! The images, combined with the disappearance time, allowed us to determine a fairly precise location, near a distinctive dark area in the Voyager images. I gave the newly discovered volcano the name Kanehekili, after a Hawaiian thunder god.

Now I was armed with two new and powerful techniques for studying Io's volcanos: I could use the infrared camera and Mauna Kea's superbly steady air to see them directly, and I could locate them by timing their disappearances behind Jupiter. I finally had my niche, and could contribute something unique to studies of my favorite world. The opportunity could not have come at a better time: my temporary job at the University of Hawaii was nearing its end, and I needed something like this to establish myself as an independent scientist. I embarked on a program to take infrared images of Io as often as the telescope time allocation committee would let me, and was lucky enough to secure a NASA research grant to fund the studies. The Io results, and my new grant money, helped me to secure a long-term job at Lowell Observatory, among the ponderosa pines of Flagstaff, Arizona. I often wonder how different my career might have been if it had been cloudy on that magical night before Christmas Eve, 1989.

Life's journey

Since that night, I and my colleagues have made more than a hundred movies of Io's occultations by Jupiter, on Mauna Kea and at Lowell. Sometimes visiting relatives join me at the telescope, continuing the tradition established in 1970. There is always more to learn, because Io is always changing. Kanehekili and Loki continue to glow brightly for our cameras, though Loki's brightness varies dramatically from season to season. Every so often these old standbys are dwarfed by brief, brilliant, eruptions that last anywhere from a few days to a few months. When we measure the temperature of these brief eruptions, we sometimes find that they are so hot, and vary so quickly, that violent "fire fountains" of molten rock may be required to explain them.

We are far from being the only Io watchers. Bill Sinton retired in 1990, but several teams, led by Jay Goguen, myself, Bob Howell at the University of Wyoming, and others, continue to watch Io's volcanos, each of us with our own unique niches. Other astronomers apply their ingenuity to observations of Io's surface composition, its tenuous sulfur dioxide atmosphere, the glowing clouds of sulfur, oxygen, and sodium that escape from Io, or to other dynamic aspects of the Jupiter system, using ground-based telescopes and the Hubble Space Telescope. Since 1996 we Earthbound observers have been joined by the Jupiter-orbiting Galileo spacecraft. Galileo can observe Io and the rest of the system in far more detail than we can from Earth, but from the Earth we can observe more frequently, and with different techniques, so our efforts are complementary. We all coordinate our work through an informal network called the "International Jupiter Watch."

Events in early 1997 give an example of how we work together. In late February 1997 Galileo took an infrared image of Io that showed that the volcano Loki was cool and quiescent, as it had been since the first Galileo observations. We had all been hoping that Loki would put on a show for Galileo but, so far, no luck. Then, just 20 days later, Christophe Dumas, working on our Io volcano team, took our first 1997 images of Io from the NASA IRTF on Mauna Kea, with the rest of us following along in Flagstaff over a computer link. Just as on our first run in 1989, Loki was blazing away, dramatically brighter than when Galileo had seen it in February. A new eruption was under way! Galileo's next observations weren't due till early April but, as luck would have it, another series of occultations of Io by Europa was beginning, and as in 1991, Jay Goguen was planning to observe many of these events, using the famous Palomar 200″ telescope near San Diego as well as the NASA Infrared Telescope on Mauna Kea. The detailed view provided by the Europa occultations showed the exact size and location of the Loki eruption as it progressed, while later Galileo images showed that the eruption had remarkably little effect on the visible appearance of Loki. By combining the Galileo data with the various ground-based programs, we will get our best chance yet to figure out what is really going on when Loki undergoes one of its spectacular eruptions. One idea we are considering is that eruptions of lava melt vast quantities of sulfur, which spread across the 150-mile expanse of the volcano, but we won't be surprised if it turns out that we're wrong.

Mauna Kea, above 40% of the Earth's distorting atmosphere, gives the best view of Io that you can get from the surface of our planet. To do better, you have to go into space, and short of going all the way to Jupiter, the Hubble Space Telescope is your best bet. After the justly celebrated December 1993 repair mission, when Space Shuttle astronauts corrected Hubble's initially flawed vision and brought it to its full potential, it provided a view of Io five or six times sharper than even Mauna Kea's best. Not surprisingly, several of us Iophiles wanted to look at Io with this wonderful instrument.

Observing with Hubble is very different from using a conventional telescope. Competition for time on Hubble is fierce: requests outnumber the time available by seven to one, and you must make a very convincing case that your program is doable and can provide important new results, before you stand a chance of getting past the time allocation committee. Once that hurdle is passed and time is allocated, you must plan every detail of the observations long in advance. Your instructions are checked and radioed up to the telescope, which does your bidding, and all you can do is sit back and hope that you hadn't overlooked some fatal flaw in your plan. It's usually several days after the observations that you get to look at your images, and there's always a moment of tense anticipation when you bring up the first image on your computer screen. Was the exposure correct? Was the image taken at the right time? And if the observation is successful (it usually is, despite your nervousness), what new discoveries await? Sometimes, you get exactly what you planned, but sometimes that first look at the images reveals something wonderfully unexpected.

It seems that our Hubble images of Io have always caught us off-balance, never quite showing us what we were looking for, but throwing up a succession of interesting surprises. Working with fellow astronomers Melissa McGrath and Paola Sartoretti of the Space Telescope Science Institute (which operates Hubble), and Alfred McEwen of the US Geological Survey, with others joining in as time progressed, I got my first Hubble images in March 1994, just after the repair mission. We hoped to see large changes in the patterns on Io's surface in the 15 years since Voyager, and perhaps also to catch some plumes of volcanic debris being blasted off the surface, as Voyager had seen. When we got the images, there were (strangely) no big changes since Voyager, and no plumes, but we did discover that parts of the surface were much more brightly colored than we had expected. So in 1995 we forgot about looking for changes and plumes, and instead took more images of Io at different wavelengths to learn more about the unusual colors. This time, however, we saw a huge surface change, bigger than any in the previous 15 years: a bright yellow spot that hadn't been there the previous year. It was the aftermath of a huge volcanic eruption at the volcano Ra (named for the Egyptian sun god), which had covered a quarter of a million square kilometers of Io in bright debris. Furthermore, we saw a hint of a plume from the volcano Pele, but it was very faint. So in July 1996 we decided to look more carefully for plumes, but the images were taken at the wrong time, so that Io was in front of Jupiter instead of having dark sky in the background. As a result, I thought, the images were scientifically useless. Still, they were visually dramatic, and I sent a copy of one of them off to the Space Telescope Science Institute, so that they could release the image to the general public. It was a couple of months before I looked at the pictures more closely, and when I did, I was amazed to find that in some of the images, because Jupiter was in the background, I could see a 400 km high plume from the volcano Pele rising off the edge of Io's disk, silhouetted against Jupiter. Ironically, if the image had been taken at the right time, with dark sky in the background, the tenuous plume would have been much harder to see. So in July 1997, we tried to see the Pele plume in front of Jupiter again, but it just wasn't there: it had switched off sometime since the previous July. Instead, we discovered a new plume where none had been seen before, and it had a quite different color from the Pele plume.

Closeup Galileo images of Io, taken before the Hubble images but sent back to Earth later, showed the new plume in spectacular detail (see Figure 6.2, color section), and identified its source—a volcano called Pillan. Later Galileo images, taken in September 1997, showed that the Pillan plume had wrought dramatic changes on Io's surface, creating a new dark spot 400 km across. With luck, our Hubble images should give us some clue about what the Pillan plume is made of, and what causes its unusual color. We wonder what surprises Io will present to us next time we point Hubble in its direction, assuming, of course, that we can convince the time allocation committee to let us take another look.

Something that always delights me about our ground-based Io work is that we base our study of extraterrestrial volcanos on the summit of a dormant terrestrial volcano, Mauna Kea, on one of the most volcanically active islands on this planet. As I drive down from the dark summit after an evening measuring Io's volcanic glow, I can often see the volcanic glow of the lavas of Kilauea, Earth's most active volcano, off in the distance to the southeast. Different planet, same process, seen in the same way: a powerful reminder of the bonds between the sister worlds of the solar system. So when I was asked to take part, in early 1995, in a series of television broadcasts which would feature our Io volcano work alongside the work of terrestrial vulcanologists working, on much closer terms, with the flowing lavas of Kilauea, it seemed very appropriate, and I jumped at the chance. The broadcasts were part of the annual "Jason Project," the brainchild of oceanographer Robert Ballard. Every year the project puts out a series of live, interactive, TV shows for school children, showing scientists at work in a particular area, in this case the Big Island of Hawaii. Every morning for two weeks, weather permitting, my wife Jane and I, joined by telescope operator Dave Griep, drove up from the dormitory at 9,000 feet to the NASA Infrared Telescope, focused on Io, and reported hourly to the TV audience on what we were up to and what we could see. On the fourth day of the broadcasts the fates smiled on us and there was that huge eruption on Io, one of the largest ever seen. (It happened to be my birthday, making this eruption one of my best-ever birthday presents.) I hope we were able to transmit our excitement to all those watching children, and that they caught something of the thrill of discovery that makes science such a uniquely exciting enterprise.

Now I have been observing Io's volcanos for eight Earth years, more than half a Jovian year of 12 Earth years. Every year Jupiter passes into a new constellation—Gemini when we discovered Kanehekili in 1989, Scorpio when the big volcano erupted on my birthday in 1995, Capricorn when Loki re-awoke in 1997. Every year prime Jupiter observing season shifts around the terrestrial calendar by one month, shifting my life with it. On a faster timescale, Io zips around Jupiter every 1.8 days, and I sometimes find myself zipping across the Pacific to keep up with it, observing a Jupiter occultation of Io from Mauna Kea and then, while Io completes one circuit of Jupiter, dashing back to Flagstaff to observe the next Jupiter occultation from Lowell Observatory. I find it deeply satisfying that my life is tied so directly to the movements of the other worlds in our solar system. We are all children of this solar system, products of a planet whose very formation was probably influenced by the gravity of the nascent Jupiter, members of a species that probably owes its existence to Jupiter's nudging of a comet or asteroid in our direction 65 million years ago, which by its impact wiped out the dinosaurs and made way for us. But few of us feel the influence of Jupiter in our daily lives (the newspaper horoscopes notwithstanding), and I feel privileged to be able to feel that cosmic connection so directly. On my way to the telescope dome, or just on my way home from a movie in the evening, I still look up at the brilliant light of Jupiter whenever it is visible. I feel an even greater sense of awe, of emotional connection, now than I felt when I watched Jupiter from my back garden in Lancashire, all those years ago. We know so much more now about the wonders contained in that bright point of light, and knowledge can only increase wonder.

7

Titan: a world seen but darkly

JONATHAN LUNINE

Jonathan Lunine. Like most of the scientist-authors in this book, Jon Lunine's fascination with *his* favorite world, Saturn's planet-sized moon Titan, began early in his career. John grew up in New York, and was educated both there—at the University of Rochester—and then at Caltech, in Pasadena, California. In 1986, shortly after completing his PhD, Jon accepted a professorship at the University of Arizona's world-renowned Lunar and Planetary Laboratory, where he has spent his career working on theoretical problems that range from Jupiter to Titan, and from the nature of comets to the constituency of stellar brown dwarfs. In his spare time, when he isn't serving on scientific steering committees, or exploring the outer solar system, Jon and his wife Cynthia, along with their son, can often be found on the mountains and rugged trails of the southwest. But the clear skies and dry, hot southwestern clines serve in stark contrast to the cold, perpetually cloud-shrouded, and perhaps rainy world called Titan that has fascinated Jon for so long. Here, Jon shares with us the story of that fascination, and the tale of how humankind has penetrated the veil surrounding Saturn's most important moon.

1 Corinthians 13

For now we see through a glass, darkly; but then face to face: now I know in part; but then shall I know even as also I am known.

Imagine a world somewhat smaller than Mars and bigger than Mercury, where the air is denser than that in this room—with a pressure about the same as at the bottom of an Earthling's swimming pool. Yet the distant Sun could never be seen by human eyes, and high noon is no brighter than twilight on Earth. The cold is so great that water is always frozen out of the atmosphere; yet the simplest organic molecule, methane, takes its place as cloud-former, rainmaker, and perhaps even the stuff of lakes or seas. Methane gas, wafted hundreds of miles above the surface of

135

this world, is cracked open by sunlight and cosmic rays; as a result, a menagerie of more complicated organics is produced, and these float down to the surface to accumulate over time. Volcanism and impacts shape the surface and provide energy to make ever more complex organic molecules in a planet-wide tapestry that is an organic chemist's dream.

This is Titan, a moon orbiting Saturn, partly revealed to humans by Voyager 1 in 1980. Titan has fascinated me for much of the last two decades, because it is the only moon in our solar system which has a dense atmosphere, and because of the atmosphere's haziness we do not know what is at the surface. Titan is a world yet to be truly explored. What awaits beneath the orange haze of its atmosphere? What exotic chemical reactions take place in its atmosphere, rich in carbon, hydrogen and nitrogen? What could Titan tell us of the early history of our own planet? I will describe how my own fascination with this mysterious world has evolved in step with international plans to explore Titan, plans which will come to fruition only after the turn of the twenty-first century.

Beginnings

My involvement in understanding the nature of Titan extends back to the close flyby of this world by Voyager 1 in 1980. But to understand why Titan was a Voyager target requires going back to the beginnings of telescopic studies of the solar system. The mid seventeenth century was a time of active observation of the classically known planet Saturn. At the turn of the century Galileo Galilei had turned his small telescope on the planet and detected what he thought were two companion globes to the main planet, these companions gradually changing in appearance. Over a generation later, in 1655, the Dutch scientist Christiaan Huygens improved the telescope to the point that he could see that the two companion bodies were in fact a ring which completely encircled Saturn. In that same set of observations, Huygens discovered a satellite of Saturn, which he named Titan. By the turn of the twentieth century the advent of telescopic apertures exceeding a meter in diameter made it possible to determine that Titan must be among the largest satellites of the solar system. Catalanian astronomer Jose Coma-Solas studied Titan's disk in 1908 and convinced himself that it seemed to darken from the center to the edge, as if he were seeing a thick atmosphere rather than a solid surface. Some 36 years later Gerard P. Kuiper used a spectrograph attached to a telescope at McDonald Observatory in West Texas to spread the wavelengths of light from Titan across a photographic plate. At certain wavelengths the light, originating from the Sun and reflected off of the disk of Titan, was dimmed. From laboratory studies those wavelengths were well-known to correspond to the molecule methane, composed of one carbon atom surrounded by four hydrogen atoms (summarized by the chemical symbology CH_4). This simplest of hydrocarbons could exist as a gas, liquid or solid at the low temperatures plausible for Titan, given its great distance from the Sun. Further study of the spectra strongly indicated the methane was a gas, but whether it was a minor component of a massive atmosphere or the sole constituent of a thin atmosphere could not be gleaned from Earth-based studies.

Voyager 1 and the atmosphere of Titan

When Voyagers 1 and 2 flew by Jupiter in 1979 I was an undergraduate at the University of Rochester, still debating whether to pursue graduate education in traditional astronomy or in the newer field of planetary sciences. I marveled at the clarity and beauty of the images of Jupiter and its four Galilean satellites. At the time I did not know that a decision was being made by the Voyager Project to direct Voyager 1 on a path that would take it very close to Saturn's moon Titan, a decision that would profoundly affect the future of the United States and European space programs. I knew little about Titan and had no inkling of the time and energy I would spend puzzling over this distant world.

On a fall evening in 1980 I stared at a TV monitor in a classroom at Caltech in Pasadena, California, where I was a recently arrived graduate student in planetary sciences. On the screen, every few minutes, another featureless picture of a portion of Titan's disk would appear.

Voyager 1 was passing only thousands of kilometers above Titan, and its cameras were programmed to take snapshots across the disk of the giant moon. Like anxious airline passengers peering through the plane's window to gain a glimpse of a foggy destination, the assembled students and faculty stared at every image, trying to find a darker patch which might signify a break in the haze. But Voyager's cameras, with the sensitivity of typical television camera tubes, could cover only the part of light's spectrum which is blocked by Titan's haze.

Had Voyager 1 been a simple photographic mission to Titan alone, its flyby of Titan would have been close to a complete scientific failure. But Voyager 1 (and its successor, Voyager 2, following some 8 months later) were amply equipped with a variety of instruments, including spectrometers that could sense both short wavelengths (ultraviolet), and the long wavelengths of infrared light, as well as a radio system that could serve to probe Titan's atmosphere in addition to calling home about the whole experience. Just a day after the featureless pictures came back I watched by closed circuit TV a press conference at JPL where the results from these other experiments were announced. The solid body of Titan turned out to be just slightly smaller than that of Ganymede, the solar system's largest moon. The surface temperature was low, 95 K (−178 Celsius), but not as low as it might have been without an atmosphere. The surface pressure of 1.5 bars found by the radio experiment is 50% larger than that at sea level on Earth.

But most intriguing was the result derived from combining information from the ultraviolet, infrared and radio experiments. The atmosphere, like Earth's, was predominantly molecular nitrogen. No water, of course, could be present at the low temperatures in Titan's atmosphere, but methane turned out to be the second-most abundant gas, perhaps 10% by number at the surface and 2% higher up. Many other molecules containing hydrogen, nitrogen and methane were found by Voyager to reside high in the atmosphere, making Titan a rich site of carbon chemistry. The starting ingredients for the chemistry were methane and nitrogen (see Figure 7.1).

This was all very interesting to me as a newly minted graduate student, soaking up all of the novel results Voyager and its seemingly heroic band of scientists and engineers could dish out. But it wasn't

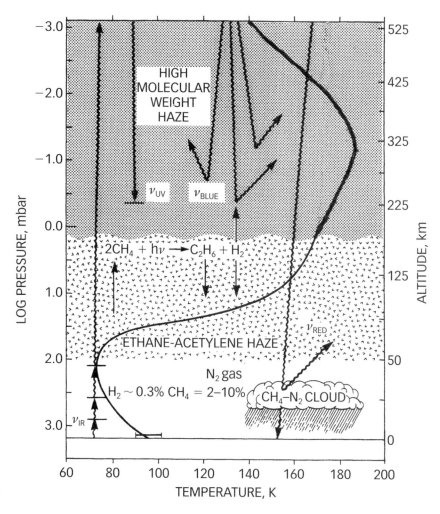

Figure 7.1

Diagram showing much of what we know of Titan's atmosphere. The curved heavy line represents the Voyager-derived temperature profile shown with altitude (right scale) and pressure (left scale in millibars; sea level pressure on Earth is 1,000 mb). Hazes occur above 50 km altitude with uncertain distribution. Light, or photons, are symbolized by the Greek letter ν; $h\nu$ symbolizes the energy from light going into the breakup of methane. The paths of red, blue and infrared light are suggested by the squiggled lines. Figure by author.

directly related to the work I had begun doing with my dissertation advisor at Caltech, David Stevenson (himself newly arrived at Caltech). I was supposed to figure out how the moons of Jupiter formed. And so, after a time out for immersion in the new discoveries of Voyager at

Saturn, I went back to pencil, paper, computer, and the moons of Jupiter.

Titan's surface

Speculations about the nature of Titan's surface were made even before Voyager 1 arrived at Titan, ranging from solid water ice with a little frozen methane trapped in pores, to vast seas of liquid methane. The Voyager 1 results permitted methane to be present as both a gas and a liquid, but the amount of methane detected in the lower atmosphere did not seem consistent with vast expanses of methane seas.

Stevenson and I began puzzling over Titan in 1981, the year after the Voyager 1 encounter. Computer models had been constructed by Yuk Yung at Caltech, by Darrell Strobel of the Naval Research Labs, and others, to determine just how methane and nitrogen act to initiate the production of the more complex organic products seen in Titan's atmosphere. Before the encounter, it was known that methane could be split apart by solar ultraviolet light to form active radicals, which would then combine with other fragments of methane and nitrogen to make more complex molecules. In the process, some of the hydrogen in the methane was freed and escaped rapidly from the weak gravity well of Titan. Thus the conversion of methane and nitrogen to more complex organics was an *irreversible* one in this view, because hydrogen escaped each time a methane molecule was broken up. The more complex computer models developed after the Voyager 1 flyby could predict how long the inventory of methane in Titan's atmosphere would last under this steady ultraviolet onslaught. The result was startling: just a few tens of millions of years, or 1% of the age of the solar system, would suffice to remove the existing methane from Titan's atmosphere.

Might we be looking at the last gasp of methane on Titan? This seemed implausible, and so the alternative was to imagine vast seas of liquid methane on the surface, feeding more methane to the atmosphere as the sunlight consumed it to make other organics. But as Carl Sagan and Stan Dermott of Cornell University pointed out in 1982, such seas had to be global, otherwise the tides raised by mighty Saturn would quickly cause Titan's orbit to be a perfect circle. Telescopic studies over hundreds of years established that Titan's orbit is in fact, slightly egg shaped or eccentric. Yet if Titan's methane seas were global, the atmosphere near the surface should contain much more methane than Voyager actually detected. This was the conundrum that increasingly occupied the attention of myself and my dissertation advisor as 1982 wore on. How could one sustain the methane without global oceans, and how could one have such oceans in contact with an atmosphere relatively "dry" of methane?

The answer lay in some work at NASA Goddard Space Flight Center by Michael Flasar, who pointed out that if the ocean were not pure methane it might coexist happily with less methane gas in the air above it. But what were the other ingredients? Stevenson and I reasoned that the products of methane chemistry high in the atmosphere, which were known by their chemical properties to form aerosols and fall to the surface, might mix with the methane in the ocean. In fact, the dominant product of methane chemistry, ethane, was a liquid at the surface temperature of Titan and would handily dissolve in the methane. This would suppress the tendency of the methane to evaporate and hence yield an atmosphere above the ocean quite consistent with Voyager data. We calculated the size and composition of such an "ethane–methane" ocean using the Voyager atmospheric data as a constraint and found it to be about kilometer deep, with enough methane to sustain the atmospheric cycle for billions of years.

Incredibly, others had overlooked the properties of ethane and hence this simple solution to the ocean dilemma. We joined forces with Yuk Yung, who had developed the atmospheric chemistry model, and in early 1983 wrote a paper for the journal *Science*. The model was quite popular, and as we found out narrowly escaped being number two in the game of research to be first with a good idea. The NASA Goddard group came to the same realization about ethane as we had, but did not get a paper together quickly enough. Though still a graduate student, I sympathized with their plight: the Jupiter work I had been doing with Stevenson at the time of the Voyager 1 encounter was scooped some months later by a research group in Japan!

The birth of Cassini and the death of Titan's global ocean

One of the perks of coming up with a model of Titan's surface that seemed to explain the Voyager data was the opportunity to tell the story to groups of more senior scientists. I did this at a workshop on satellites at Cornell in Spring 1993, and was thrilled to have Carl Sagan step up to the stage after my talk and make remarks which clearly indicated his approval. However, a presentation I gave later that year at JPL to a joint committee of US and European scientists was more significant. They were considering what sort of follow-on missions to the Voyagers at Saturn should be conducted, and an orbital survey seemed appropriate. But should an entry probe be targeted at Saturn just as NASA's Galileo probe was to enter the Jupiter atmosphere, or should it be aimed at Titan? One to each was too expensive, and other bodies of senior scientists had weighed in favor of Titan. But what might the probe find at Titan? Should it be designed to float? What key measurements at the surface ought to be done? I was given my five minutes in the Sun to pontificate to these great scientists.

The Titan work opened some doors. I completed further studies of how an ocean and atmosphere would evolve over time on Titan, bundled that up with some more technical studies of water ice and presented the dissertation to the Caltech faculty in early summer 1994. They blessed it, signed the papers needed for a PhD certificate, and pointed me east out of Pasadena. The destination was the University of Arizona, where (except for a short stint at UCLA) I have been ever since. Just a year after arriving at Arizona I was on my way to Europe courtesy of Don Hunten, one of my new bosses and a member of the National Academy of Sciences. European interest in designing a probe for Titan's atmosphere, to be ferried to Saturn by the Americans, was picking up steam. In Europe two meetings were held to explore what could be done by such a mission in the Saturnian system. My fascination with what was to become the Cassini mission to Saturn and the Huygens probe to Titan was tremendously increased on this visit.

By the late 1980s I was on the faculty at Arizona and doing studies with graduate students of the evolution of Titan's surface and atmosphere through time. There was little new data to confirm or challenge the ethane–methane ocean story. The first chance to do so came in 1988 when Caltech professor Duane Muhleman and colleagues succeeded (after other groups had failed) in sending a radar beacon from Earth and receiving the reflection from Titan. That reflection was much too intense to have come from a body covered in liquid hydrocarbons. Subsequent radar experiments revealed that the average returned signal was smaller than seen in the earliest experiments, indicating that different parts of Titan's surface had different compositions, but still out of bounds of a global hydrocarbon ocean. At least some, and perhaps much of the surface was a solid material, with ice (bearing some amount of less reflective material) being a prime candidate. However, the radar lacked the resolution to see any detail on Titan and hence, as bold and successful an experiment as it was, could only hint at the true nature of the surface.

Progress toward a sharper picture was coming at the same time from observations just beyond the wavelengths of visible light, in the infrared. It was known that Titan's obscuring haze became more transparent to light of progressively longer wavelengths, but a second obscuring factor was the atmospheric methane itself, which soaked up light over broad swaths of wavelength. University of Hawaii planetologist Tobias Owen realized that, by looking carefully at selected wavelengths where this absorption was weakest, it might be possible to peer through the fog of Titan's atmosphere and get some hint of the surface. This approach bore fruit from ground-based telescopes around 1990, which detected color variations from different hemispheres of Titan as the moon was observed in different phases of its orbit around Saturn. Then, in 1994, a postdoctoral fellow of mine, Ralph Lorenz, teamed up with Peter Smith and several other colleagues from the University of Arizona and York University in Canada to use the newly repaired Hubble Space Telescope in these specially transparent infrared wavelengths. Though not a particularly big telescope, Hubble's position above the blur of Earth's atmosphere allowed it to see details on Titan's surface down to a couple of hundred kilometers linear distance across Titan's surface. Though hardly impressive by flyby spacecraft standards (a picture of Earth at that sharpness would miss London and barely detect the greater New York area), it did reveal a surface with bright and dark mottlings. Most impressive was an Australia-sized bright region, located on the face which had returned the strongest radar echoes. Was this a plateau of water ice sufficiently high to trigger methane rains, which in turn washed dark organics into lowlands?

Subsequent infrared studies with Hubble and some specially adapted ground-based telescopes went in two directions. Some compared the light coming from Titan through several of the methane "windows" to try to determine what the surface was made of, a difficult proposition given that the atmosphere had influence even in these relatively transparent parts of the spectrum. Others used new techniques on giant telescopes which helped correct or remove the distorting influence of our own atmosphere, in attempts to better Hubble in making images. Figure 7.2 (see color section) shows one of the successes—an image of Titan's surface made on the giant 10-meter Keck I telescope in Hawaii, within an infrared methane window, by Bruce McIntosh and colleagues of the Lawrence Livermore Laboratories. The image reveals a greater complexity of light and dark areas than did Hubble. As one might expect with increasing sharpness, progressively smaller areas of greater brightness and more profound darkness appear. And some of the dark areas are non-reflective enough to be lakes or seas of methane and ethane.

The scientific controversy surrounding the surface of Titan was heating up. The absence of broad areas of very dark material seemed to rule out the global ocean, while still allowing more limited lakes and seas of liquid hydrocarbons. A graduate student of mine, William Sears, and in a simultaneous but separate effort Sagan and Dermott, showed that seas confined to closed basins (such as craters made by impacts) would not raise tides sufficient to circularize Titan's orbit. But such seas would not be great enough in volume to store methane for billions of years of future chemistry, nor would they hold enough ethane to account for billions of years of past chemistry. Stevenson suggested that perhaps the ocean is under the surface, stored in a very porous layer of ice. Alternatively, the record of billions of years of chemistry may have been destroyed by large impacts associated with nearby Hyperion, a satellite which appears to have recently broken up.

And then, again, we must face the possibility that we see Titan at an unusual time in its history, when it has methane in its atmosphere: perhaps this happens only every so often. I, my students, and colleagues around the world still struggle with these issues, but our zeal for doing so would not be nearly so great were we not to have any prospect of finding the answers.

That prospect became much greater in 1990, when the US space agency NASA and the European Space Agency ESA agreed to begin development of a large vehicle to orbit Saturn and a saucer-shaped probe to descend through Titan's atmosphere to its surface. From 1986 to 1990 various committees, including one I chaired for NASA, debated the worth, form and cost of the project. But the potential for spectacular science along with international collaboration were big selling points. I like to think that the controversy surrounding Titan's surface played a role. In response to a NASA solicitation for instruments and scientists to be a part of the mission, I applied for the job of "Interdisciplinary Scientist" for Titan's atmosphere and surface. Such a position would give me access to data from multiple experiments, and allow me to participate in mission planning. I was in Friedrichschafen Germany, for a scientific conference in September 1990, when the selections for the ESA Huygens probe were announced. I was thrilled to find I had been selected, along with a wonderful payload that really could tell us what is the nature of Titan's surface.

The Orbiter carries a solid state camera which can peer into one of the infrared methane windows to Titan's surface, an infrared camera and spectrometer, and a radar capable of taking images of the surface. In addition, another spectrometer on the orbiter will measure heat coming from Titan, a radio system will sound Titan's atmosphere á la Voyager and measure carefully the shape and hence interior structure of Titan, and other devices will determine how Titan interacts with the magnetic bubble known as Saturn's magnetosphere. Though this essay focuses on Titan, it is important to realize that the mission of the Orbiter includes an in-depth survey of Saturn's rings, atmosphere, magnetosphere and other icy moons besides Titan.

The Huygens probe carries six instruments including a combination camera and infrared spectrometer (designed here at the University of

Arizona), two chemical devices to sample the gases and particles of Titan's atmosphere, as well as temperature, pressure and lightning monitors. A radar and acoustic sounder will probe the surface before impact in ways other than taking pictures, a precise radio timer will measure how winds carry the probe across the surface, and instruments will determine how soft or hard the surface is at impact. And, yes, the probe will float. After relaying data to the orbiter during a two and a half hour descent, the probe can transmit for thirty more minutes from the surface before the orbiter "sets" over the probe's local horizon. Powered by lithium batteries to save weight and cost, the probe will wind down and die before the next pass at Titan by the Orbiter some weeks later. However, hundreds of images and large amounts of other data will be transmitted from the probe during its three-hour mission. After relaying probe data back to Earth, the Orbiter will settle into a four-year tour of the Saturn system. Shaping its orbit through frequent close passes by Titan, the Orbiter will sail through various parts of the Saturn system. And as it passes by Titan to alter course, the instruments on the Orbiter will build up a global view of Titan to compliment the "up close and personal" exploration of the Huygens probe. Even partial success of the instruments will tell us what lies beneath Titan's hazy atmosphere. But more than that, Cassini-Huygens will allow us to attack questions of the long-term history of Titan's atmosphere and surface. I summarize two of the questions which I find most fascinating, both of which have potentially sobering implications for understanding the early history of our own planet and the origin of its biosphere of which we are part.

Titan and the faint early Sun

If Cassini-Huygens fails to find evidence of broad methane-bearing seas, then how is the photochemistry sustained over billions of years on Titan? If Stevenson's suggestion of underground reservoirs is not right, one possibility is that Titan is a "Jekyll and Hyde" planet, with epochs rich in methane and other times bereft of the hydrocarbon. What might Titan be like during the lean years? Since methane is one of the gases which warms Titan's surface by retaining infrared light, how cold could the surface get without its modest greenhouse effect? Simulations of Titan's climate by myself, postdoc Ralph Lorenz and NASA Ames scientist Chris McKay suggest two answers, related to the astrophysical claim that the Sun has warmed about 30% over the history of the solar system. In recent times, a loss of methane modestly cools the surface about 10 degrees, not enough to force nitrogen to condense and drop out. Thus a methane-less atmosphere today would look similar to the actual atmosphere, though it would be more transparent to the eye since hydrocarbon hazes would be absent. But early in Titan's history, the fainter Sun would have kept the giant moon more precariously close to the point at which the nitrogen itself could condense to form clouds, and rain out. With methane depleted a crisis threshold is crossed in our models. In our picture of a Titan on which methane comes and goes, early epochs of methane depletion could have led to atmospheric collapse.

How might Cassini-Huygens find evidence of such ancient episodes of atmospheric collapse? We know that Venus, a thick-atmosphered planet, has large impact craters but not small ones. The smaller bolides break up in the dense atmosphere and disperse as small particles. On airless or thin-atmosphered bodies such as Earth's Moon or Mars, craters of all sizes are present, the smaller ones actually being most numerous. Titan's present atmosphere is thick enough to shield small impacts to almost the same extent as Venus does. If its present atmosphere has been sustained throughout history, we expect the probe camera and orbiter cameras to witness only large craters on Titan—if craters exist at all. If instead the cameras pick up a distribution of small and large craters, then we would have a strong hint that, early in its history, Titan's atmosphere waxed and waned with its supply of methane. But more than that, since a faint Sun is required to bring Titan's atmosphere to the threshold of collapse, we will have our first direct evidence of a faint early Sun. Such evidence would be profound, since both early Earth and early Mars show evidence of being *warmer*— not colder—than they are today. This apparent contradiction of the astrophysical understanding of the Sun and our knowledge of the ancient climates of Venus and Mars remains a thorn in the side of studies of the early evolution of Earth, Mars and the emergence of life on these worlds. It is fitting, I think, that the explorations of Titan by Cassini-Huygens might shed a new light—however dimly—on this long-standing problem.

Titan and the origin of life

And what about life? For all the superb research done in laboratories, we do not yet know how life on Earth came to be. To find other worlds where the events of life's origins may be playing out today, on a natural tapestry much larger and longer-lived than any Earth-bound labora-tory, is a long-standing goal of planetary exploration. And Titan is certainly a target for that kind of research.

So cold that liquid water is only a transient product of volcanism or impacts, Titan almost certainly is not the home of life today. But its carbon chemical cycles may constitute a natural laboratory for replay-ing some of the steps leading to life. Though more hydrogen rich than the Earth likely ever was, Titan is probably the closest analog to the Earth before life that we can study in the solar system. Reactions involving life's key elements—carbon, nitrogen, hydrogen and oxygen, have been ongoing for much longer and on a much larger scale than can ever be hoped for in the laboratory.

We know that life is abundant on Earth, and has played key roles in our planet's evolution; much of the evidence for how that life began has been erased. We suspect that Pluto, comets and other outer solar system bodies probably retain the original inventory of organics from the beginning, but have undergone little evolution toward the forma-tion of life, because of the extreme cold and airless nature of those objects. Then there are three objects—Mars, Europa, Titan—which may have undergone some amount of organic chemical evolution: almost to the threshold of life, perhaps, on Titan, possibly beyond it on Mars and maybe Europa. While Mars once had liquid water, and Europa may have it today, Titan has lacked this essential ingredient for most or all of its history; how far can one go toward the origin of life without it? Cassini-Huygens can at least tell us whether Titan is the right place to ask such questions, and what tools we might invent to conduct follow-on explorations. Regardless of the answer, it is safe to say that neither of the mission's namesakes might have imagined the exotic nature and enduring mysteries of this moon to which the last three centuries of observation have given witness.

Coda

Every so often you could see faintly the head of an alligator as it bobbed to the surface, scattering fish from its immediate vicinity. Such a sight, by itself, is no reason to stand at the edge of the Banana River, outside Cape Canaveral Florida, at 3 a.m. in the morning. But this was October 15, 1997, and 8 miles away a Titan IVB Air Force rocket was being readied to send a new spacecraft to Saturn and Titan. If launch was successful, Cassini-Huygens would be on its way to peer beneath the haze of Titan to tell us, once and for all, what this moon's surface was really like. And I was there, with my wife and young son, to see it happen. Seven years of work by thousands of people, including myself, sat on top of the booster. Cuts to the NASA budget twice threatened the mission, which was saved largely through the presence of Europe on the mission. And, finally, in the months before launch, a noisy campaign was forged by some groups to cancel Cassini because of its use of plutonium as a heat source to generate electricity on the spacecraft, as had been done for all outer planet missions before Cassini. Yet all these threats seemed suddenly to pale before the prospect of a rocket launch, the most hazardous part of any space mission.

Two days earlier, bad weather and minor technical problems turned the countdown into a tense dress rehearsal. This morning, Wednesday, milestones were checked off with clockwork precision over the loud-speaker. I felt calm and even the sounds of the alligators seemed soothing. My wife Cynthia assured me that we would launch today. Then, almost too soon came the final backward count from five to zero, a rich orange flame appeared to shoot outward from the rocket's base, and Earth's emissary to Saturn and Titan moved smartly away from its launch pad. The rocket passed through a cumulus cloud, illuminating it from within, and then arced over through the night sky. We could hear and feel the thunder build and then fade, and the last we saw of the mighty namesake of Saturn's mysterious moon was the dim blue light of its first stage engine pushing its payload higher and higher above the Atlantic. Later we would learn that the launch was so accurate that little or none of the first course correction would be needed, and early checkout of the Huygens Probe revealed no problems.

This book may be out of print by the time Cassini-Huygens arrives at Saturn in July, 2004, with deployment of the probe at Titan later that year. But I would hazard a guess that Titan's surface will still be a mystery by that time. Should all go well for Cassini-Huygens, that year and the four following it will see discoveries beneath the orange haze of Titan not dreamt of today.

8

Triton is doomed

BILL McKINNON

Bill McKinnon. Triton may be doomed, but Bill McKinnon is blessed. Bill, a planetary theorist, was born in Canada and raised in New York and Louisiana (what a combination!). Bill took his undergraduate education at MIT and his graduate studies at Caltech. Since then, he has focused much of his attention on the outer solar system, but he has also been influential in work on planetary impacts across the solar system, the tectonics of Mercury and Venus, and he's even had his hand in some observations of radio stars, of all things! Still, Triton has always been one of Bill's research passions. After sampling life on all three US coasts, east, west, and gulf, Bill settled in the American heartland, teaching and researching at Washington University in Saint Louis. And what kind of person is he? A husband, a father, an artist (if ever science were art!), and a music lover. In the essay below he tells us why Triton so fascinates him and, in doing so, gives us all an entertaining and insightful look into perhaps the most bizzare moon our solar system has to offer.

TRITON IS DOOMED. So proclaimed a column in the science section of *Time* magazine on October 14, 1966. It went on to describe the recently published work of one Thomas McCord, a Caltech graduate student whose "two-year, computer-aided investigation" showed that Triton, the largest of Neptune's then known moons was "doomed to smash into Neptune in a cataclysmic collision" in "as little as 10 million years." Ten million years! That's nothing compared with the solar system's age of 4.5 billion years, or the Sun's estimated total main sequence life span of 11 billion years. By the cosmic clock, 10 million years is practically tomorrow: If the Sun and planets were destined to live only a human lifetime, Triton would meet its unmaker in just one month.

This was heady stuff for a seventh grader whose passion was space exploration in general, and the planets in particular. I grew up in a time when rapid scientific and technological progress was the order of the day, if not an American birthright. Jet planes, atomic energy, rockets to the Moon, color television, stereo. I didn't need to be told that science was important—it was cool. With science, one could, by being really smart and using the power of one's own mind, change history, change everyone's lives forever. And besides, I was good at science and math, and I think it is natural if you are good at something, whether it's math or playing guitar or shooting hoops, to want to see how far you can go.

I was especially intrigued with stories of scientific *mysteries*, such as how John Couch Adams in England and Urbain Le Verrier in France could simultaneously and independently deduce, using Newton's Theory of Gravitation, the existence and position of a then unknown planet from irregularities in the orbit of the recently discovered Uranus, or how Edward Jenner figured out that milkmaids who caught cowpox, a relatively mild disease in humans, were somehow conferred an immunity from the dreaded smallpox, which led to the first vaccine in human history and an end to one of humankind's oldest scourges.

I suppose my interests could have turned in a number of directions, growing up in and around New York City. Perhaps if my parents had listened to my childhood entreaties and taken my siblings and me to the Statue of Liberty or the Empire State Building, I would have ended up a sculptor or an architect. For whatever reason, what my father did do was take us, when we went into Manhattan, to the American Museum of Natural History and adjoining Haydn Planetarium. And you know what that means: dinosaurs and outer space!

I suppose most children, certainly most boys, enter into dino-mania and space-craziness at some point. But as much as I loved dinosaurs, it seemed to me that all the big expeditions had been mounted, the major fossils found, and that most of what could be figured out about ancient life had already been figured out (or would only be learned very slowly). In more ways than one it seemed that paleontology was a dead field, and that that most profound of mysteries, the riddle of the disappearance of the dinosaurs, was not, I thought, destined ever to be solved.[1]

[1] I was wrong on this point, but I later took great satisfaction when the extinction of the dinosaurs became an important topic of study in planetary science, through the miracle of impact cratering.

Space, on the other hand, was clearly the field of the future. In elementary school we were marched to the auditorium to watch lift-offs; the TV was wheeled into the cafeteria so we could keep tabs on John Glenn's three orbits of Earth (I still remember the bad chow mein that day); class was interrupted so we all could watch Ranger 9 live-broadcast pictures of the surface of the approaching Moon, right down to the last crash-ful instant.

Even when the schools had lost interest, I got myself up at dawn to watch Walter Cronkite's coverage of the first launch ever, unmanned but quite successful, of the Saturn V, *the* Moon rocket. But the Apollo program and the Moon were not really my principal interests; I wanted to explore, to know about *the planets*. Here were real mysteries. Was there or had there ever been life on Mars? What lay behind the opaque clouds of Venus? What did an asteroid, any asteroid, look like? And what about all those mysterious moons, off in deep space, circling the giant planets?

Outer space is incomparably vast, and the planets of our solar system are only a miniscule part of it. I realized early on that the distant stars, the immeasurably more distant galaxies, the creation of the Universe in the Big Bang, and so on, although fascinating, were for me forever out of reach in space or time. As long as Einstein's special relativity held true, and the speed of light unsurpassable, I was unlikely to see even the closest stars close up. No matter how much scientific progress was made on the origin and evolution of stars, galaxies, and the Universe as a whole, I would never truly *know*. But the planets were essentially in our own cosmic backyard, demonstrably reachable by our machines, if not ourselves. I knew I had been born in a special time, what Carl Sagan later called The Golden Age of Planetary Exploration. So I chose the planets (or perhaps, they chose me).

Cosmic thing

Adams and Le Verrier's mystery planet turned out to be Neptune, of course, and Triton is its largest moon. Discovered only $2\frac{1}{2}$ weeks after Neptune in 1846, Triton has long been considered quite a puzzle on its own. Its orbit is cockeyed—it circles Neptune at a substantial incline to the planet's equator, unlike all the other major satellites of the giant planets (Jupiter, Saturn, Uranus, and Neptune), or our Moon or the moons of Mars for that matter. What's more, it moves backwards in its orbit, or "retrograde," with respect to Neptune's rotation, something that "normal" moons simply don't do (see Figure 8.1).

All this clearly intrigued Thomas McCord in the mid-1960s, and what his calculations showed was that Triton could have had a much more distended orbit in the past—an orbit that could have extended to the limit of Neptune's gravitational reach. Looked at from the other end, what this also implied was that Triton's peculiar orbital geometry

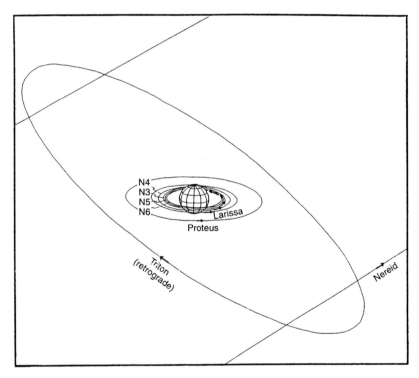

Figure 8.1

Orbits of Neptune's family of satellites. Shown is a perspective view along a line of sight inclined 18° to Neptune's equatorial plane. The innermost satellites are all relatively small and were not discovered until Voyager 2 passed through the Neptune system. They orbit in Neptune's equatorial plane, while much more massive Triton circles outside them in an inclined, retrograde orbit. All the satellites have virtually circular orbits except for Nereid. The apparent crossing of Nereid's and Triton's orbits is an artifact of the projection. Figure from Kargel (1997), in *Encyclopedia of Planetary Sciences*, edited by J. H. Shirley and R. W. Fairbridge, Chapman & Hall, London.

could be due to its being a *captured* moon, captured from elsewhere in the solar system. This surely wasn't a regular moon at all! Being captured from elsewhere, its initial orbital geometry about Neptune would have been a matter of chance, so having an inclined, retrograde orbit is just as possible as any other orientation. McCord's calculations showed that after Triton was captured (by a mechanism he did not explicitly discuss) its vastly extended orbit would have shrunk and been circularized under the influence of the surface tides that Neptune and Triton would have raised on each other.

Almost as an afterthought, the same calculations showed that, because Triton's orbit was retrograde, these same tides should be presently causing Triton's orbit to continue to sink toward Neptune, and thus, someday in the future, Triton will ultimately collide with Neptune or be torn apart to form a massive ring system, such as the one that circles Saturn today.

This last prediction is what caught the *Time* science writer's eye, and what fired my imagination. It was also just what I needed for one of my oral reports in 7th-grade science class, so I carefully copied the illustrations from the magazine showing Triton's orbital track, inserting a few extra time steps for clarity, onto a pungent mimeograph master. And so, in the fall of 1966, I gave my first scientific paper, such as it was, on Triton.

Planetary karma

What goes around does seem to come around. In the spring of 1979 I was a graduate student at Caltech, taking a course on planetary dynamics from Professor Peter Goldreich, a world-renowned celestial mechanician. Years earlier, he had (among others) inspired and taught Tom McCord the basics of what Tom needed to model Triton's orbital evolution. Now Goldreich was talking about the planet Pluto, and how some had hypothesized that Pluto was an escaped satellite of Neptune. Goldreich noted that this wasn't true, but he didn't say exactly why. Actually, there were no scientific papers that said it wasn't true, and several that said it was; in fact, the idea was certainly what one read (and sometimes still does) in astronomy textbooks. I filled out a mental index card to look into this someday. That someday actually came, thanks to Triton.

For more than a century following Triton's discovery, it had been maddeningly difficult to find out anything concrete about this moon. It was too far away for astronomers to see a disk even using the big 200-inch telescope atop Palomar Mountain. So we didn't know how big it was, and seeing surface detail was out of the question. Nor did we know how massive it was. Triton's distance and orbital period about Neptune were used, with Kepler's third law, to determine the mass of Neptune and Triton together, but the much smaller mass of Triton could not be separated from that of Neptune by this technique. Given this dilemma, one had to try something else to get at Triton's mass. It was known that the effect of Triton's mass would show up as a subtle wobble in Neptune's position, but the detection of Neptune's tiny wobble would be a dicey proposition, at best. Nevertheless, astronomers claimed to have done just that, in the 1930s and 1940s. Measurements from photographic plates indicated a surprisingly large mass for Triton.

The problem in determining Triton's size was that, besides being so far away, Triton is not all that bright an object in the sky—not much brighter than Pluto really. We didn't know how big Pluto was either back then, but educated guesses about how reflective the satellite's surface might be, combined with the measurement of Triton's brightness, put some limits on Triton's size. And these limits combined with the early estimates of the mass implied that Triton might have the density of solid gold! Well this would be great if it were true, but planets and satellites, at least the ones we know, don't have densities that high! What this really meant was that we didn't know diddlely about Triton. It nevertheless continued to be noted that Triton might be the largest moon in the solar system (because of its supposed large mass), which at least fed the impression that Triton could be one of the solar system's most important pieces of real estate, gold or no gold.

This situation began to change in the late 1970s and early 1980s. Ground-based infrared astronomy had progressed to the point where it was finally possible to identify chemical species on Triton's surface. What were detected, by research teams led by Dr. Dale Cruikshank, then of the University of Hawaii, were not precious metals, but extremely volatile ices. Methane was the first to be identified (in 1979), which was interpreted as meaning Triton had an atmosphere, and which made it only the second satellite then known to possess one (Saturn's Titan being the other; see Jon Lunine's essay: chapter 7).

In 1983, a related, and more electrifying discovery was reported. A very faint, single spectral line was identified as an absorption due to nitrogen, N_2. The absorption of infrared light by nitrogen is so intrinsically weak, though, that no reasonable atmospheric thickness could account for the dip in Triton's spectra. But the Sun's infrared rays reflecting through a mass of condensed nitrogen, either solid or liquid, would show such an absorption. Cruikshank and company reasoned that the temperatures required to freeze nitrogen to a solid were possible, but unlikely, and so concluded that Triton was probably covered by substantial bodies of *liquid* N_2. This was a fantastic scene to imagine: exposed oceans on a moon orbiting Neptune, deep in the frigid outer solar system. Some found this too fantastic. The spectral feature was weak and had never been detected on a body in space before, but Cruikshank had a great track record of pulling spectral rabbits out of telescopic hats. Dale knew his instruments and could walk the line between inspired deduction and overinterpreted noise. There was nitrogen there, and perhaps it was even in liquid form. Triton was no longer simply intriguing; it had become downright weird.

And as the saying goes, when the going gets weird, the weird turn pro. It was time, in the early 1980s, to begin to seriously reconsider the nature of Triton. I say reconsider because the old theory about Pluto being an escaped satellite of Neptune involved Triton as well. This goes back to Pluto's discovery in 1930. When it was determined that Pluto's orbit follows a rather eccentric elliptical course, one that periodically takes it closer to the Sun than Neptune is (as Pluto has been recently, ending March, 1999), a young British astronomer named Ray Lyttleton took this to mean that at some point in the past the orbits of the two planets might have physically intersected, which in turn suggested that Pluto could have at one point been bound to Neptune. From this, Lyttleton proposed that Pluto and Triton formed originally as normal, regular satellites of Neptune, that is, close to Neptune and in its equatorial plane, like the regular satellites of the other giant planets. Some slow change, possibly outward tidal evolution of the satellites' orbits (such as affects our Moon), brought Pluto's and Triton's orbits too close together for celestial comfort. Perturbations in both orbits built up until one or more close encounters completely destabilized the system: formerly prograde Triton was hurled into a tilted, retrograde orbit, while Pluto was completely ejected from the Neptune system, becoming the planet we know and love.

Theories are subject to further corroboration, or disproof. And the 1949 discovery of a second Neptune moon, diminutive Nereid, in a distant, extremely eccentric orbit (albeit not a retrograde one; see Figure 8.1) certainly spoke of a former cataclysm in the Neptune satellite system. Nereid's unusual orbit could be accommodated in Lyttleton's framework: it was another original moon, one whose orbit was also changed substantially, just not one sent retrograde or ejected. It was perhaps a bit of a stretch, but nothing *too* serious.

The first serious challenge to Lyttleton's theory occurred in the mid-1960s, when electronic computers were used to calculate the orbits of the planets for the first time. Their results showed a curious thing. As far back as the computers could accurately calculate, Pluto never came close to Neptune. That is, when looking at the solar system from above so that the orbits of the planets are seen in projection, Pluto's orbit indeed crosses that of Neptune (see Figure 8.2, top). In three dimensions, however, Pluto's orbit *never* came close to touching Neptune's, because Pluto's orbit is inclined to Neptune's, and no matter how Pluto's orbit oscillated over time, it never lost the particular geometric arrangement that served to maximize the distance between Pluto and Neptune as they both circled the distant Sun (see Figure 8.2, bottom).

What these first computer simulations had actually discovered, although the full intricacy of it would not be revealed for another 20 years, was a "gravitational resonance" between Neptune and Pluto (actually a series of nested resonances, like a Russian Matryoshka doll), in which Pluto's orbit is always phase locked to Neptune in just the right way. For example, Pluto's orbital period (averaged over time) is precisely 50% greater than that of Neptune. In any event, those computer calculations could not say whether Pluto had always been in this

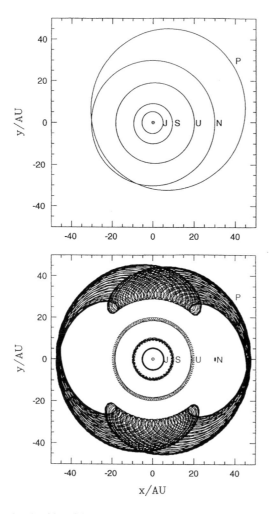

Figure 8.2

(Top) Orbits of the giant planets (J, S, U, N) and Pluto (P) projected onto the ecliptic plane (the Earth's orbital plane); the scale is given in astronomical units (AU), the distance from the Earth to the Sun. Officially, from January 21, 1979 to March 14, 1999, Pluto was the eighth planet from the Sun. (Bottom) The same projection, but this time showing the actual calculated positions of the planets over a 40,000 year period, fixing the Sun at the center and rotating with Neptune's average (or mean) motion. In this uniformly rotating frame, Neptune's position oscillates slightly, whereas the curlicue motions of Jupiter, Saturn, and Uranus are due to the slight eccentricities and inclinations of their orbits. Despite the pronounced spirograph-like trajectory that Pluto takes with respect to Neptune, it never comes close to it; one can see that Pluto passes closer to Uranus than it ever gets to Neptune. Pluto is stable in this resonant configuration, caught in Neptune's gravitational thrall. Figure from Malhotra and Williams (1997), in *Pluto and Charon*, edited by S. A. Stern and D. J. Tholen, University of Arizona Press, Tucson.

resonance, because one couldn't hope to cover all 4.5 billion years of solar system history with 1960s computers, but neither was there any reason to think that Pluto could be ejected from Neptune and end up in as special a dynamical niche as the resonance.

In the summer of 1982, as a postdoctoral research associate (post-doc) at the the University of Arizona in Tucson, I was looking for some new mystery to think about, to research, to dig into. I wanted something different than projects connected to my PhD thesis work, which had concerned the mechanics of impact crater formation. I had already moved from craters on the Moon and the inner planets to impacts on the icy satellites of Jupiter and Saturn (newly revealed by the Voyager encounters of 1979–1981), to research on these moons themselves; now I was preparing to leap frog it all. Laying out my mental index cards, I found the mysteries of Triton (and Pluto) compelling. I also felt the peculiar pleasure that hardly anybody was seriously working on or thinking about these distant bodies. Certainly, from the point of view of their origin and early evolution, I had it all to myself. On the other hand, this also meant that nobody cared (yet). Of course, if in the end I didn't find out anything important or at least interesting, it would have all been a waste of time, for I had no real research plan, no new instrument, no detailed hypothesis to test (yet). On the other hand, it has always seemed to me that when you venture out into the unknown, in science as in all other things, marvels await.

I had always been dissatisfied with Lyttleton's escaped satellite hypothesis, so I decided to look at that first, and I had one new fact to put in the mix: I knew Pluto's mass, or rather, the mass of Pluto and its moon Charon combined. Charon had only been discovered in 1978, but its mere existence allowed Kepler's third law to again prove itself useful. The mass of Pluto–Charon is not much, about 20% of the mass of our Moon, but more importantly, using any reasonable estimate for Triton's size and density (based on its brightness and likely constitution—rock and ice, not gold), the mass of Pluto–Charon was less than the mass of Triton. It was a simple matter of physics (conservation of angular momentum and mechanical energy and all that) to show that petite Pluto simply could not, through any conceivable set of gravitational interactions, reverse the orbit of more massive Triton, should they both have started out as normal, regular satellites of Neptune.

Now, it may be a good thing to trash an old hypothesis that has outlived its usefulness, but it is more satisfying to contribute something positive, to place a brick in the wall of science. To this end I wanted to at least offer an origin "story" to replace Lyttleton's. He had started Triton and Pluto as satellites of Neptune, and then sent one into solar orbit. A more logical, alternative hypothesis was not hard to imagine; it was essentially just sitting there, waiting for someone to apply physical reasoning to the early history—the assembly—of the outer solar system. Our modern view of this assembly is that the planets accreted from countless lesser worlds, which we call planetesimals, or if large and distinguished enough, protoplanets. Triton and Pluto comfortably fit into the category of large planetesimal or protoplanet. Essentially, what I proposed is that Lyttleton's hypothesis be reversed: Triton and Pluto start out as small planets—*satellites of the Sun*, and then one, Triton, becomes captured by Neptune.

In this framework, Triton and Pluto would not in themselves be unusual bodies, as in Lyttleton's scheme, but rather the survivors of a once much vaster ensemble of similar worlds orbiting near and beyond Neptune. They are with us today because each entered into a special dynamical relationship with Neptune, relationships that protected them from further major harm (accretion into a planet or ejection from planetary region). Pluto became locked into the resonance with Neptune, whereas Triton became Neptune's satellite.

A large outer solar system planetesimal population also provides a ready explanation for Pluto's relatively large moon; Charon could be the result of a catastrophic collision between proto-Pluto and another precursor body, a scenario which I proposed at a meeting in 1982. In the resulting published research paper, I also proposed that Triton could have been captured by Neptune by means of gas drag (aerobraking), when it passed through the extended gaseous envelope or disk that probably surrounded the young Neptune.

I have always been pleased that my early "giant impact" idea for Pluto and Charon predated the famous 1984 Kona Conference on the Origin of the Moon, where the giant impact hypothesis rose up to become the governing paradigm of lunar origin, but bemused that I was not bold enough in my thinking to apply this idea to Triton as well (and me, a guy trained in impact mechanics). That honor went to a team led by Peter Goldreich (there he goes again), who proposed in 1989 that Triton was captured when it collided with an original (and no doubt smaller) moon of Neptune. This economical little idea would have pleased William of Occam, because gas drag capture is a tricky mechanism to get to work right (and I have spent quite some research time on that very task). All I can say is that I wish I had thought of it first![2]

High tide, hot moon

The best part about working on Triton was that something fantastic really did come up. Starting with the hypothesis that Triton really was captured, it was natural to dig up McCord's (now) old paper and reexamine his orbital evolution equations. This was useful because by the 1980s we had knowledge of Neptune's internal constitution unavailable to McCord. It was possible to make a better defined (and hopefully more realistic) set of calculations. A fundamental point is that as Triton's orbit changed from a very extended ellipse just after capture to the nearly perfect circle it is today (see Figure 8.3), there must have been a corresponding, and large, change to the mechanical (kinetic plus gravitational potential) energy of the system. Numerically, it would be just about equal to the energy it would take to fire some monster rocket on Triton today and take the entire moon out of Neptune orbit. But in Triton's case the "engine" is not a rocket, but tides. And the process runs backwards; Triton gets closer to Neptune, so all this energy is apparently lost from the system. I say apparently because the total energy is conserved. In Triton's case, as the tides rise and fall, the lost mechanical energy reappears as *heat* inside Triton.

The amount of heat pumped into Triton as its orbit evolved could have been huge. First, Triton's ices would melt and the rock would fall

---

[2] On the other hand, the recent discovery (in October 1997) of one or two distant, retrograde, and presumably captured satellites of Uranus may revive interest in the gas drag capture hypothesis for Triton and other satellites.

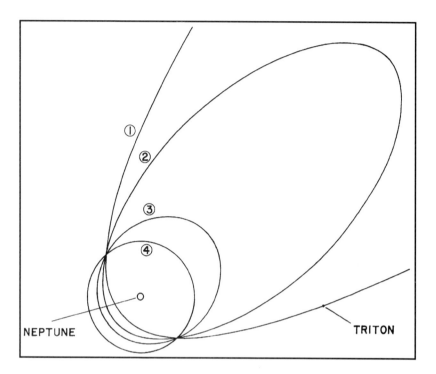

Figure 8.3

Schematic tidal evolution of Triton's orbit, subsequent to capture by Neptune. A very extended elliptical orbit (1) shrinks and circularizes (2 through 4). In doing so, Triton courses through much of the space close to Neptune, and would have profoundly disturbed, if not destroyed, any original satellite system. Only satellites quite close to Neptune could have survived unmolested (see Figure 8.1). Nereid, Neptune's outermost moon, may be a perturbed remnant of this original system. Triton would have been spectacularly heated and melted during this orbital evolution. Figure from McKinnon et al. (1995), in *Neptune and Triton*, edited by D. P. Cruikshank, University of Arizona Press, Tucson.

to the center, forming a core; then this core would itself melt, forming a metallic inner core and a silicate outer core. Both would be mostly if not entirely molten, and the contact between the liquid rock (magma) and the overlying deep water ocean would be a scene of more or less continuous violent boiling. Triton would be so hot that its most volatile ices, such as methane and nitrogen, would be raised in a thick atmosphere around it. Indeed, Triton's surface temperature and atmospheric conditions should have been more determined by the tidal heat flowing out of it than by the feeble light of the Sun, which at Neptune's distance is a thousand times less intense than on Earth.

Why wasn't this picked up on sooner, you might ask? One can't fault McCord, because tidal heating was not the subject of his 1966 paper, nor could one have made a very convincing case back then based on what was known about Neptune and Triton. Fundamentally, I think that, until the Voyager era, the concept of planets or satellites being heated by tidal dissipation was an obscure one. This all changed when Voyager 1 sailed by Jupiter's Io in 1979, and we have since gone on to discover that heating due to tides is or has been important in many icy moons and the planet Mercury too. But Triton is the body that, at least theoretically, could have been heated more severely than all the others. It would, however, have been a one-shot deal. Unlike John Spencer's Io, which is continuously tidally heated (because of its orbital resonances with neighboring Europa and Ganymede) and volcanically erupting right now (while you read this!), once Triton's orbit circularized it would be basically all over, and from that point on the satellite would follow a more prosaic evolutionary path.

An epoch of tidal heating would surely have left its mark on Triton. Even if the epoch was billions of years in the past, and the satellite long cooled, we might expect to see unusual surface geologies, perhaps unlike anywhere else in the solar system. Perhaps the chemistry of the surface and atmosphere would be exotic (one could always hope), with organic compounds as a distinct possibility. And these were not idle speculations with no way to find out, for we (the human race) were on our way to Triton. Voyager 2, outbound from Saturn, was scheduled—after a brief pit stop at Uranus, and if the spacecraft remained healthy—to arrive at the Neptune system in late August, 1989.

Satellite on the edge of forever

The Voyager encounters with the giant planets were unique in the annals of planetary exploration, and all had a somewhat Star-Trekky feel. After years of interplanetary travel, it almost seemed as if each spacecraft would "drop out of warp" near its intended planetary target—with each giant planet and its family of moons resembling a miniature solar system to be explored. And it really would appear to be "where no one had gone before."[3]

The trajectory of each encounter would not only take the spacecraft by the planet and ring(s) in question, but by as many moons as geometry would allow. The data were taken and the images snapped and radioed back, largely immediately, to the Earth. Immediately in the case of Neptune meant taking over four hours to reach Earth, traveling at the speed of light. Picked up by the supersensitive antennae of NASA's Deep Space Network, the image data were then processed by automated routines into video pictures, which were in turn transmitted to the abundant monitors at the Jet Propulsion Laboratory (JPL) in Pasadena, the Caltech campus, and any other place that was linked up in those pre-internet days. All anyone had to do was put down his or her coffee and watch: every few minutes frame replaced frame, each one a picture of something that had never, ever, been seen before in human history.

The Neptune–Triton encounter would be the sixth and last for the Voyagers, following two encounters at Jupiter, two at Saturn, and one at Uranus. It was the culmination of an incredibly productive and successful space project (all possible because of RTGs (radioisotope thermoelectric generators), by the way). I attended all of them, starting with the Jupiter encounters in 1979, which were scientific and exploratory events so rich and spectacular that they gave a big shot of reanimating juice to my graduate career. The Neptune–Triton encounter was particularly anticipated, by scientists, by the science press, and by the science-interested public. This was due at least in part to the efforts that a small band of researchers (folks such as Dale Cruikshank, Jon Lunine, Alan Stern, the editor of this collection, and myself) had put in earlier in the decade, helping to take research on Triton and Pluto from obscurity to the planetary forefront. In the end, people did care, something I found enormously gratifying.

Most of Voyager 2's best pictures and other data would be returned

[3] Technically, Jupiter and Saturn had previously been visited by the Pioneer spacecraft, but the differences between the quality and quantity of data returned by these relatively unsophisticated machines and the Voyagers that followed are enormous, especially as concerns the satellites.

during the week-or-so long near-encounter with Neptune, but at a full year out the resolution of Voyager 2's camera exceeded that of any Earth-based telescope. From then on, everything would be new.

One of the first things the images determined was Triton's true size. Finally. By August 22, when Voyager was three days from closest approach to Neptune, the front page of *The New York Times* was able to announce "Profile of Neptune's Main Moon: Small, Bright, Cold, and It's Pink." So Triton turned out not to be as large as Jupiter's Ganymede or Saturn's Titan after all. Rather, it was a modest 2,700 km in radius, smaller than our own Moon, though plenty larger than all of the so-called mid-sized icy satellites of Saturn and Uranus.

To account for its brightness in telescopes, Triton's surface would have to be more reflective than previously thought, and reflecting more sunlight means the surface is colder. Triton's surface temperature was eventually measured by Voyager to be an astonishingly low 38 Kelvin (−391 degrees F), the lowest ever measured on a solar system body, too cold for liquid nitrogen, but just right for solid nitrogen ice. Triton's visible color had long been described as "red" by astronomers, but the technical red of astronomers is not the intense red of Gaugin. Now that we knew how reflective (icy) Triton was, the papers could legitimately call it pink—Triton, the designer moon!

Voyager's approach to the Neptune system gave it a good view of Triton's southern hemisphere, which was mostly in sunlight because it was southern summer there (what a concept). What the approach images showed was a vast icy polar cap extending up almost to Triton's equator (see Figure 8.4, color section). Subtle marks and streaks tantalized, but nothing specific could be made out. Voyager's trajectory would carry it to within 5,000 km of Neptune's cloud tops; skimming across the blue giant's northern polar region just after midnight on August 25, 1989 (PDT), the spacecraft's path would be bent by Neptune's gravity so that the plucky robot would rendezvous with Triton some 5 hours later. So much data would be taken in so little time that much of it would have to be stored on the onboard tape recorder for later playback. The camera exposures would be long, for the imaging system was designed for the light levels at Jupiter and Saturn. The spacecraft would be moving so fast with respect to Triton that the entire spacecraft would have to pivot or nod as it took each frame in order to prevent image smear (a bit like sports photography).

The highest-resolution images came down later that day (see Figure 8.5), and they revealed a surface of breathtaking diversity. This was no ordinary icy satellite, with stacks upon stacks of impact craters over here, some fractures over there, and perchance an ancient icy volcanic plain or two. Triton's surface is unlike anything seen before, on any planet or satellite. The quarter of Triton imaged at high resolution contains three if not four different and unique terrain types, which can be seen in a computer mosaic assembled from the individual high-resolution images (see Figure 8.6, color section).

At the equator on the east (right) side of the mosaic lie volcanic plains, but these are not just smooth expanses of material; many volcanic vents and collapse craters can be identified, many surrounded

Figure 8.5

All eyes are on the monitors in one of the imaging team areas as the high-resolution Triton sequence comes in. Smiling in the foreground are, from the left, Drs. Edward Stone (Voyager Project Scientist, and now Director of JPL) and Bradford Smith (Imaging Team Leader). The bearded man in the upper center with his hands his pockets is University of Hawaii astronomer Thomas McCord. I was also in the room, but not in the picture. Photograph courtesy of David Morrison.

by fresh icy flows. Moving to the west and towards the terminator (the shadow line), the vents are replaced by vast sheets of some icy material. Exceptionally smooth quasi-circular regions have been interpreted by some as frozen lakes; I see the edges of these lakes as formed primarily by sublimation and scarp retreat. Regardless, the volcanic materials are relatively fluid. In terms of possible lava chemistries, both ammonia-water and methanol-water eruptions have been suggested.

Moving farther to the west, we encounter the amazing cantaloupe terrain. Seen close up in Figure 8.7 (color section), this terrain is characterized by an organized pattern of non-circular dimples (now called cavi). And believe it or not, there exists a geological parallel on Earth: domes of salt that rise buoyantly through overlying denser sedimentary rock layers and break through to the surface. Salt would not be a very sensible component for Triton's icy crust, but the evidence for instability and overturn in the cantaloupe terrain suggests an unusual layering of different kinds of volatile ices or their response to heating from below. This is not the only possible explanation, but it is the one offered by Paul Schenk of the Lunar and Planetary Institute in Houston, a former PhD student of mine, so you might say I'm partial to it (and I do think it's the best explanation).

Most of Triton's southern hemisphere is given over to the southern polar "cap," which in Figure 8.6 appears discontinuous at the edges. In Figure 8.7 the cap can be seen encroaching on the cantaloupe terrain from the top. The polar ice appears to bury the preexisting terrain, so it probably really can be considered an ice cap of some 1 kilometer maximum thickness. The most mobile of the ices on Triton is nitrogen, followed by methane, and the southern cap (and northern cap, if it exists) are the major nitrogen reservoirs on Triton's surface. Nitrogen gas (with a trace of methane) subliming from the caps and elsewhere on Triton supplies the bulk of Triton's atmosphere. Because of the extremely frigid conditions on Triton, this atmosphere is quite thin; a surface pressure of 16 microbars was measured during the Voyager encounter. Despite a surface pressure equivalent to that at 75 to 80 kilometers altitude on the Earth, Triton's atmosphere is substantial enough to support thin clouds and hazes. What's more, Triton's unusual orbit translates into extreme seasonal variations in solar heating. Over periods of hundreds of years (because of Neptune's stately orbital pace), Triton's atmospheric pressure rises and falls by a factor of 10 in each direction with respect to its 1989 value.

And if this were not enough, already, Voyager discovered *plumes* of gas and icy particles rising up from the polar terrain to heights of about 8 km before being sheared by the prevailing Tritonian winds. The many dark streaks on the polar cap (Figure 8.6) turn out to be the fallout from previous plumes. What powers these plumes is unknown. The two leading suggestions invoke the concentration of sunlight, and heating from below, respectively.

All of the terrains I have described are very lightly cratered, and even though we don't know the absolute flux of bombarding comets at Neptune, the implication is that Triton is quite young geologically. All in all, the picture Voyager 2 painted of Triton was one of a dynamic and presently geologically active world. And just think of what might exist on the unseen other three-quarters of the satellite!

Can anything Voyager discovered be laid at the doorstep of massive tidal heating? It is very tempting to do so when one looks at all the unusual geology, the volcanically active regions, and even some peculiar chemistry—CO_2 ice was discovered on Triton in 1991 by Cruikshank and colleagues and is not seen on any other icy satellite. But to make a conclusive case is more difficult. After all, if Triton's capture occurred early in Neptune's history (billions of years ago), which is the most likely time given that there were many more loose protoplanets back then, should we expect to see the surface bubbling along today? Some have suggested that it was Triton's final orbital circularization and tidal heating that was delayed until more recently, but in my opinion this is special pleading, and requires an unusual set of circumstances and behaviors.

No, it is unlikely that Triton is running on tidal energy today, which means that the ices in its surface layers and deeper mantle must be volatile and mobile enough to respond to the modest heat now flowing out of the satellite's interior. The arrangement and chemistry of Triton's icy surface layers may have been determined during the era of intense tidal heating, but we can't tell yet. We need more sophisticated models of Triton's geological evolution, with or without tidal heating, and more and better knowledge of Triton's surface geology and chemistry to provide a clearer answer.

A key thing the Voyager encounter with Triton did confirm was Triton's link with Pluto. The close passage of the spacecraft to Triton allowed the satellite's mass to be well determined from the change in the spacecraft's velocity. Triton's density turns out to be very close to twice that of water, and very close to Pluto's density as well.[4] This density, in bodies the size of Triton or Pluto, implies that both are composed of about 70% by weight rock and 30% by weight ice. This is also close to the composition that I and others predicted for bodies that form in solar orbit far from the Sun, but it is conspicuously more rocky than observed in regular icy satellites. Triton and Pluto are now fairly well accepted as sibling bodies, similar in size, density, and composition (Pluto has also been found to possess a thin nitrogen atmosphere and N_2 and CH_4 ice on its surface). A close up look at Pluto, something NASA has been studying but has not committed to, would be a key test of the tidal heating hypothesis for Triton, because Pluto would not have suffered the same fate.

And something else amazing has come up since the Voyager encounter. Pluto and Triton are not alone. Astronomers, starting in 1992, have been identifying new solar system bodies near and beyond Neptune's orbit by using the most sensitive electronic (CCD) cameras ever developed with the biggest new telescopes at the best observing sites in the world (such as Mauna Kea). More than 60 have been discovered so far, and although they are nowhere near as large as Triton or Pluto, they are not tiny either. The largest is estimated to be almost 800 km across (and there probably are tiny ones, we just can't see them). Their orbits are spread out beyond Neptune in a sort of belt, named after the famed mid-century Dutch-American planetary astronomer Gerard Kuiper, who predicted the objects' existence nearly 50 years ago. What's more, a number of these Kuiper belt objects are caught in the same gravitational resonance that Pluto is; David Jewitt,

[4] Determining Pluto's density required determining Pluto's size, which has now been done by timing the passage of a star behind Pluto from various vantage points on the Earth, and by timing the passages of Pluto and its moon in front of one another.

the University of Hawaii astronomer who co-discovered the first Kuiper belt object (besides, if you will, Pluto), and who with his collaborators is responsible for more Kuiper belt object discoveries than anyone else, refers to this particular subset of the Kuiper belt objects as Plutinos— little Plutos.

Kuiper belt objects too small to see (or at least see easily) may be vastly more numerous. The Kuiper belt is now thought to be the second great solar system reservoir, after the Oort cloud, of Paul Weissman's beloved comets.

What do the Kuiper belt objects themselves represent? They are the remnants of the planetesimal swarm that Pluto, and the object that we believe collided with Pluto to make Charon, and Triton grew up in. And it's just the tip of the iceberg. For example, in 1991 Alan Stern looked at the likelihood of Triton being captured by Neptune and Pluto's moon forming in a massive collision, and determined that there had to have once been hundreds if not thousands of Triton-sized bodies plying the spaceways out near and beyond Neptune. Some may still survive, and await discovery. We'll just have to wait, and see.

The once and future Triton

After all of this, you may be wondering whether Triton really is doomed. Technically, the answer is yes. The tides that Triton raises on Neptune are inevitably dragging Triton closer and closer in. But we have a much better idea of the relevant properties of Neptune and Triton, after Voyager, than in the 1960s. And to be fair, McCord presented a host of possibilities, from which the *Time* science writer pulled the most exciting one. Redoing the calculation, we find rather than having only 10 million years left, Triton is likely to take more than 10 billion years to spiral in to Neptune. This is more time than it will take for the Sun to swell into a red giant star and boil away all the Earth's air and water. So while the Earth is broiling to a cinder, Triton may become a balmy, aquatic paradise. Certainly a strange twist for an ill-fated world.

My own hopes for Triton is that we'll return some day, with a more sophisticated exploratory craft. I don't think this will happen anytime soon, but just as with our return to Jupiter with Galileo and to Saturn with Cassini, the marvelous Neptune system calls out for a more systematic approach, by which I mean a Neptune orbiter, something that can study the planet, its rings and smaller satellites, and especially Triton, for an extended period of time (years as opposed to Voyager's days), and in great detail. For example the Voyagers did not carry instruments capable of determining the *composition* of the various geological units they saw, something that can be done, and in doing eliminate an entire layer of speculation. I am banking on techological advances that will lower the cost of exploring the outer solar system, and am comforted by the knowedge that in about 40 years, Triton will be nearing its equinox (southern autumn), when all of Triton will be illuminated over the course of a Tritonian day (5.9 Earth days). The longer we wait, at least until then, the more of Triton we'll be able to see.

On one level, of course, the essential mystery of Triton has been solved. It has moved from obscure point of light to a revealed world in a single human lifetime. There are still major puzzles to solve, but what would really "make my day" would not actually be a return to Triton.

At this point I would much rather get to Pluto for the very first time, which is the most scientifically justifiable course. It is also the most financially sensible, because a rendezvous and flyby of the Pluto–Charon system could do all the prime science one would want, and would clearly be cheaper than a Neptune orbiter. And it could go on, deep into the Kuiper belt, visiting one or more of the strange new worldlets, feeling for the edge of our planetary system. This gets at, I think, the ultimate importance of all this work on Triton and its kin. We planetary scientists, we happy few, are engaged in understanding the construction of the solar system and its component parts, and yet right now we do not know where our planetary system ends. And it is, ultimately, the exploration of new worlds that provides an intangible vitality to our science, a rush of adrenaline and endorphins, or maybe just a sense of pure amazement, that everyone on Earth can share. Life is so short. What are we waiting for?